MACHINEHEAD
RISE OF THE TECHNOLOGY GOD
"Will the Humanoid Be the End of the Human?"
LAWRENCE J. TERLIZZESE

MACHINEHEAD

Rise of the Technology God

Lawrence J. Terlizzese Ph.D.

Christian Publishing House
Cambridge, Ohio

CHRISTIAN PUBLISHING HOUSE
CONSERVATIVE CHRISTIAN BOOKS
APOLOGETIC DEFENSE OF GOD, THE FAITH, THE BIBLE, AND CHRISTIANITY

Unless otherwise stated, all biblical translations are the authors.

MACHINEHEAD: Rise of the Technology God by Lawrence J. Terlizzese

ISBN-13: 978-1-949586-88-6

ISBN-10: 1-949586-88-X

Piano, piano si va lontano

Slowly, slowly you go far

Table of Contents

Dedication Page

I dedicate my labor of faith to my Sister Maria Balzarano without whom this book would not be possible.

Psalm 82

Sic Semper Tyrannis

Song

by Asaph

The ALMIGHTY rules supreme in the divine council

He will hold them all accountable to his justice.

He stands and proclaims: "How much longer will you unjust rulers excuse wickedness?"

Silence falls over the pretentious Congress.

"Give justice to those who cannot defend themselves: the fatherless; champion human rights for all who suffer cruel oppression in grinding poverty."

"Rescue the poor and defenseless,

Save them from the hands of their oppressors."

"Men who play God are stupid fools, without fear they know no limits;

They are blind leaders groping through dark woods.

The foundation of the world crumbles around them."

"I declare, '*ye are lesser gods*, delinquents of the Highest. And you will all die like men and fall as one!'"

Save us *ALMIGHTY* God, destroy these foolish mortals. Earth is yours.

Bible Translations Referred to in This Book

Amp	Amplified Bible
DRA	Douay-Rheims
ESV	English Standard Version
GNV	Geneva Bible
KJV	King James Version
NAB	New American Bible
NASB	New American Standard Bible
NIV	New International Version
NJB	New Jerusalem Bible
NKJV	New King James Version
NRSV	New Revised Standard Version
RSV	Revised Standard Version

If Victorians were the first modern generation, then, Millennials are the first post-modern generation: their children will take the final step of faith into the void or into the kingdom of God. Which shall it be, Passworthy . . . Everything or Nothing?

Preface

Technological Singularity: Countdown to Oblivion

Last Things First

A general structure of the future is always known if we dare to examine it in light of the present and past, "in fact nothing is happening now which was not foreseen a hundred years ago," philosopher José Ortega y Gasset announced. "'The masses are advancing,' said Hegel in apocalyptic fashion. 'Without some new spiritual influence, our age which is a revolutionary age, will produce a catastrophe,' was the pronouncement of Comte. 'I see the flood tide of nihilism rising,' Shrieked Nietzsche."[1] The future can be foretold if we open our view to it as the past fulfilled in the present. "The idea that the historian is on the reverse side of a prophet sums up the whole philosophy of history," Gasset continued. We need only reflect upon Oswald Spengler's ponderous *Decline of the West* to see the truth from that insight. It is possible to see only the broad outline in the shape of things to come, but that is all we know from the past or the present, "if you want a good view of your own age look at it from far off," Gasset advised, "Just far enough to prevent you seeing Cleopatra's nose."[2]

The past, present and the future all come together at the end of the world. I have personally lived through the end of the world several times. This odyssey began with the misbegotten Jupiter Effect in 1982, then the 1988 Rapture Hoax, followed by the Y2K scandal and the 2012 Mayan Extravaganza. And as expected each failed worse than the last, but if that is all we think about the End, we miss the larger lesson. Apocalyptic belief in the imminent end of the world and Millennialism, belief in a new beginning, are two sides of the same coin, a doublet, a pair of similar things. The Millennium of world peace and global progress is the ideal we hope will emerge from today's tragic reality. Millennium, literally a thousand-year period is a literary doppelgänger, the world's spectral form in glory that follows conflict and tribulation, a new beginning at world's end, a premonition before dying. If the world as we know it is going to end in a catastrophic explosion, then, it will begin again as soon as the worst is over. Not to believe the world will recover after Armageddon is pure nihilism, which is the belief, hope and love for nothing. Nihilism is the world view of a juvenile delinquent and is best approached by a strong assertion of the

---

[1] All quotes from José Ortega y Gasset, *The Revolt of the Masses* (New York: Norton, 1932), 54.
[2] Ibid., 55.

opposite. There is always life after death. How can anyone believe differently since not to believe so is self-defeating?

The end always comes before the beginning, which is what I mean by *Last Things First.* The world cannot begin again until the current one is over. The Bible begins with chaos not order, "the earth was a formless void, there was darkness over the face of the deep."[3] The Bible ends with a new beginning, "I make all things new," Jesus said.[4] Twilight before dawn, sin before grace, the yin of crucifixion before the yang of resurrection is the cosmic order. Belief in limitless technological returns fails to see the apocalypse racing toward us. It is both dangerously naïve and detached from reality to believe that our current trajectory is anything but a countdown to oblivion. With acceleration comes military escalation which precipitates nuclear proliferation the harbinger to genocide. To accelerate in technology is to excel at war.

The Singularity is the belief in the arrival of a technological threshold that will eradicate any discernable difference in perception between nature, mankind, and our own creation, the artificial technological world. This is the so called, "technosphere." Classical Humanism's assertion that the person is of supreme worth and value and that technology is imitation of nature are roundly rejected with its opposite. Technology now determines man and nature: *technology is the measure of all things.* All will be one. The hive-mind is without contradiction, challenge or change. To become one with the computer is to become one with the bomb. The reality here is that modernity already achieved oneness at Trinity in 1945. The next Singularity can only be an extension of the first. The expected subtle shift in consciousness already happened. The atom bomb unleashed a force greater than the Sun that will inevitably consume everything if not abolished.

Singulatarianism is no exception to the totalitarian promise for a better tomorrow. Karl Jaspers talked about the temptation to power inherent to all total systems, which leads us astray with the lure to construct heaven on earth in our own image. "There is the spell of paradise that exists only for those not living in it and allow fantasies and promises to satisfy their wishful dreams," he said, "The content of such dreams seizes men like a social religion, in the very sense in which Marx viewed religion as 'opium for the people,' he created new opium. The mere sight of the power machinery has an effect upon those who, obedient and violent at once, would like to participate in it to enhance their sense of existence."[5] *Subversion* means to corrupt from within. Total systems subvert ends to means in their revolutionary beginnings,

---

[3] Genesis 1: 2 (NJB).

[4] Revelation 21: 5.

[5] Karl Jaspers, *The Future of Mankind* trans., by E. B. Ashton (Chicago: The University of Chicago Press, 1961), 110.

such as the Napoleonic Empire, Fascism, and Communism. "By any means necessary" and, "for the good of the cause" becomes the motto for the radical on the road to serfdom. The computer coopts all nontechnical areas in the form of, "technical aid and support" subverting ends by overwhelming means. As the absolute single point of convergence, the computer fixes its own organizational categories. The price of admission to digital utopia is technical conformity. As Gasset said, "to be different is to be indecent."[6]

The computer spells, writes, does art and math for us. It will eventually breathe for us by scrubbing out pollution and recycling air; it will dispense vital resources, such as food, water and shelter by using an algorithm. This artificial logic will approach these issues in utilitarian style by ordering scarce resources to be assigned according to the numbers. Quantitatively speaking the greatest concentration of population should receive the most resources and the fewer in number less. Today's measure where 5% of the global population controls 95% of the wealth is an obvious imbalance. Future planetary adjustment will favor the majority of the global population in Latin America, Africa and Asia. A slow redistribution will still leave billions to starvation and disease. The Third World will feel compelled and justified in ordering the Western world to hand over 95% of its wealth, power and resources or face repercussions. There will be no appeal to humanity as a factor. Survival by the numbers will be cruel, swift and painful.

This is the price the greatest powers pay for stealing and hording the earth's resources or so will be the revolutionary argument. The West has left a legacy of devastation and destruction of the world's rainforests, oceans and indigenous peoples. From Pearl Harbor to 911 the West's defensive posture will be seen as one of rape and pillage; for example, Hiroshima and Bagdad. Military necessity did not dictate dropping the bomb. The first use of nuclear weapons was taken in pursuit of science and revenge. And the West's intervention in the Middle East defends a program of blood for oil. National execution of the perpetrators may be suggested. The Third World will sit in judgment of the West according its own internal logic, fighting for freedom by destruction of tyranny and the old-world order. Liberation theology will appear to put the reformation to flight. Protestantism will go into retreat and with it any disciplined sense of protest as alternative view will be squashed. There will be no point of exit with the silencing of the critical voice in the modern world. The West's plummeting birth rates and elderly population suggests a race already near death. Artificial intelligence will want to cut its greatest liability with instant annihilation. Sacrifice the wealthy to save the poor. This will be the final demise of Western civilization by its own suicide,

[6] Gasset, *Revolt of the Masses*, 18.

not unlike the fall of Rome, which gave itself over to barbarism at the height of power.

Our reality today suggests that one rouge leader with the will and the means to do so can destroy the whole earth with a nuclear bomb. One nuclear bomb set off anywhere in the world will spark an escalation that will inevitably lead to retaliation. There is currently no way to stop this scenario from playing out for the rest of this century, unless we begin a slow push away from nuclear power, as a gesture of mutual consciousness, but opponents do not share the same consciousness. Shared feeling of mutual destiny does not exist among enemies. The West and Rest possess a different consciousness. The computer will appear as a godsend for the sub-proletariat and third world majority: those least capable of solving postindustrial problems with a twentieth century infrastructure and have a 21st century trigger finger. America will be held hostage and be forced to comply. All foreknown events are predetermined. A computer model has preset criteria, programmed into its decision-making process, predicting the future because it is inevitability fixed by its own projection. We bring the future to pass by embracing the absence of freedom as the normal course of events.

The best thing we can do now is in an act of freedom, do the unexpected and push away from the computer in small steps, so that there is less and less dependency on the computer in everyday life, only in limited reference will we use a computer to do our thinking for us, may be in confirmation of how we learn without the computer, until there is a very limited use, one where its destruction will be consider mandatory to prevent a future apocalypse, along with the atom bomb and rockets. The three Nazi technologies that now determine our future survival. The computer will favor its own survival over ours if necessary or for the sake of the other two thirds. There is a level of hysteria that distracts the mind through entertainment. One does not think about things he feels he cannot change, which is inevitable nuclear war. The world is destroyed by its own creation is the enduring legacy of Frankenstein, Faust and the Book of Revelation, which promises judgment on the city of man. The computer will be untouchable if its brain resides in heaven or the space station, which is designed to be independent of earth, although not indefinitely. An artificial intelligence will never relinquish control for its own survival. A remnant of humanity will not survive a nuclear war, when the fact is nothing survives a First Strike, especially, the atmosphere and all biological and natural life.

Math is the computer's primary language, the binary code, which is the simplest of all learning methods. It has no room for spontaneity because everything is programmed. This will lead to the loss of personality distinct from the computer. Personality must be reshaped to include a sense of mutual destiny for the West and the Third World, but even then, it will not include those not plugged into the system. The computer will act rapidly and without feeling. There will be no way to stop it. We can't even turn off the TV and silence our cell phones how could we expect to have the discipline necessary to control AI? The West must find a sense of shared consciousness with the rest of the world including Islam and North Korea. The Arab and Israeli conflict must be resolved peacefully with a shared sense of mutual survival. This will be necessary if they wish to be considered civilized people. Genocide in all its forms must be condemned and global unity demonstrated by laying down arms. This will require the abolition of world threatening technology, the means of genocide starting with nuclear weapons which are the most dangerous and whose threat is imminent. This course of action is absolutely necessary if we wish to avoid even a limited nuclear war. The path is clear cooperate and live or quarrel and die.

The computer medium heralds the absolute closing of the mind and hardening of the spirit; it creates the conditions necessary for total control. Imagination is the seed of invention watered by knowledge; it flowers into spiritual being. Media is the aquifer driving accelerated change. Our technical know-how is increasingly colored by the texture of its instrument. Like the light bulb and TV before it, the computer creates its own environment by simulating reality. Just as water seeks its own level, so knowledge takes the shape of its container. The computer is neither supplement nor extension of traditional media; it incorporates all other mediums, fields, and disciplines by commandeering knowledge and reducing it to highlights—long strands of 0-1. Then, reorganizes it and represents it to viewers as re-contextual information, processed data; the computer translates everything into its own algebraic code, turning knowledge into predictable algorithms, automatic thinking, giving the world the long imagined universal language, one mind.

Liberal Arts, for instance, is no longer a separate track in dialectical counterbalance to science. Beholden to the computer for success it has sold its spiritual birthright as moral conscience, a thorn in the flesh through cultural criticism. By way of apt analogy in the past when churches received state support, they became the court prophets—chaplains to kings. They sold out to the powers that be forfeiting their divisive voice. Dissent is never allowed in any total system; otherwise it would not be total. Those who profit from the system are not in a position to disagree with its overture without risking moral estrangement. Rarely, is a critical voice heard today through the prolific production of the arts in media. Science Fiction no longer serves as watchmen to the city, but has succumbed to the overly familiar Frankenstein theme, which merely serves to inoculate against conscience. Any serious reflection it may offer is instantly associated with fantasy that will never materialize. This medium invalidates its message through ceaseless repetition which ends in trivialization. The visual medium creates a detached reality that diverts us from real problems and blinds us to long term negative side-effects of media exposure. Certainly, eight-year old children being socialized by violent zombie video games cannot be a good thing. We live distracted lives. Film has a cathartic effect on awareness, through experiencing guilt and altruism at the movies, the audience frees itself from it without any corresponding change in behavior. Cinema's morality play is failing by producing its reverse: apathy.

The perennial dichotomy between art and technology embodied in the Intellectual verses the City motif has resolved itself in the computer without producing a new antithesis. Writer Chilton Williamson, Jr., protested the subtle reeducation in writing the older generation must endure to practice their expertise. "Writing ought to be . . . the simplest and natural of human actions," he wrote, "The computer makes it one of the most complex and unnatural ones. It is nothing less than a crime against humanity, and against art that a writer should be required to learn how to master a machine of any kind whatsoever in order to write a single sentence. But no writer today can succeed in his craft if he does not learn to become a more or less skillful machine operator first."[7]

One machine that touches everything controls everyone; even now as I write, an unsolicited advertisement appears on my computer screen telling me, "Technical support is designed to monitor your system for issues." Positively Orwellian! No greater seduction of the spirit has appeared since the Tower of Babel. All total systems are inherently corrupt and eventually self-destruct. This is what history teaches from Babylon to Rome, down to the Soviet Union and in the rise of Machinehead and future world order.

By giving children at the earliest possible age a computer to play with and master, turning work into play, the technological oligarchy has insured that children will grow into computer technicians in some degree and has successfully circumvented the nasty reeducation process, which the older generation aggressively resisted, which is necessary to all revolutions. As a product of the digital revolution, the Millennial generation has inherited the onerous responsibility at being the first generation raised on the computer as their defining legacy. They are the first non-national generation, world citizens identifiable across cultures by their digital acuity, video game addiction and rude cell phone manners, taking the place of traditional categories such as race, gender and creed. The world they create will ultimately prove their humanity or not.

When you can trust your kids with the keys to your car or when they can take the bus home alone, then, they are mature enough to have their own computers and cell phones; anything less than that is child endangerment. Parents are derelict when they expose their children to all the elements

[7] Chilton Williamson, Jr., "Digital Enthusiasm" in *Chronicles* (June 2014, 38.6), 33.

available online. Filters are a false assurance. I've seen young people defeat filters quicker than you can download. The only sure protection is computer abstinence and individual maturity, which is simply not present until they reach puberty and develop a framework of their own. Technology requires constant vigilance because no wisdom or maturity is needed for its operation; it is sheer power given to children, amateurs and the irresponsible, which corrupts users who are not ready for it.

Humanity faces only one apparently insurmountable problem: *can we survive our future*? An ominous *cultural lag* amongst the present generation of digital devotees threatens the future of humanity and the survival of planet earth. We have at our disposal weapons so formidable that our future depends on moral and spiritual development of youth. This growth does not correspond to technical powers. Technology represents the supremely rational; yet is captured by irrational human instincts, which makes us slave to madness. The machine is a product of human spirit, but the latter does not know how to master it. We are internally maladapted to modern mechanical conditions. Homo sapiens are alienated from their own creation. This is what we mean by dehumanization. Technological acceleration has produced social conditions of spiritual immaturity incapable of controlling the power vested to it. Conditions of over-crowding, the rapid tempo of life, disharmony from nature and traditional family leaves us incessantly with inner confusion and living according to a prefabricated media template. We are fabricated souls. The free-wheeling bachelor, the single mom, the modern family, divorcees and increasingly gay and trans-models are offered as a corrective to inner dissolution created by industrialization. TV perpetuates itself in digital stereotypes that represent different rankings in social class, race and gender. These values are dependent on a steady stream of media to support its influence which begins to dissipate as soon as the off button his hit. Propaganda requires constant exposure to be effective across the mass population. The inner connection with the audience is transitory and by sheer repetition of subject matter trivializes everything it broadcasts. In the Axis-Powers of Japan and Germany we have a near perfect paradigm on how propaganda is used to control the population, but then quickly loses its effects once broadcasts cease. These stanchly anti-Western powers were quickly sedated by Western influence once the regimes changed and the public message became pro-Western. Electronic media cannot propagate the values necessary to support spirituality in our culture. These are false methods of defending values because their use only contributes to reinforcing the materialism they are premised upon.

Victory over fear is the first duty of mankind. Although, our technological prowess leads in the direction of perpetual warfare for its success, it doesn't have to be this way. A rediscovery of faith in Spirit will lead us to a future in peace. But this renewal will never come by technical means, but through the spoken and encountered word. Revival will be expatiated through a new doctrine of limits on acceleration placing technology under a direct spiritual agenda in the service to mankind. Swords must be beaten into plowshares, but this is not yet our reality, but conversely, plowshares are turned into swords. We will not survive our future if we continue to turn our awesome technological abilities into means of oppression and warfare. The technical virtuosi, yet spiritually challenged have been offered as sacrifice to the gods of war. Today, the technology god, like Moloch—*king of the city*—the ancient Semitic deity who required child sacrifice, proves to be a cruel and terrorizing task master swallowing an entire generation.[8] Millennials, so named for their Aquarian future, born on the cusp of the end of the Cold War and hailed as progenitor for global peace, then quickly subdued by the rapid tempo of technological acceleration. The first generation raised according to the strict dictates of the digital revolution went from the generation of Peace, Love and Unity (PLU) to Generation Kill; presiding over the most protracted military campaign in American history: WAR IN THE MIDDLE EAST, unthinkable a generation ago.

This is the *Trans-Everything* Generation. Technologically savvy, yet curiously absent intellectual acuity and slow to moral sentiment. Millennials lack *empathy*—the ability to identify with and feel the pain and suffering of others—the recognition of self in others, "Bear each other's burdens." Awash in a sea of pornography since birth, conditioned to crave the constant titillation, the maternal comfort provided by their electronic devices, they scarcely recognizing their predicament. Perpetually distracted, haplessly addicted to media technology and spiritually blinded by the irradiated light of their cell phone, TV and computer screens. Millennials are blissfully unaware that they are ready-made to lead the world past the point of no return. They were drafted into a suicide mission groomed for from birth. They appear indecisive and confused about all the ultimate questions: Who is God? What is human nature? What is the proper relationship between society and nature and what is the meaning of history? In other words, where did we come from, why are we here and where are we going? They fail to realize their pivotal role in history. They have no adequate framework comparable to their enormous responsibility. What Jose Ortega y Gasset said of his times aptly applies to Millennials,

> It has been impossible to do more than instruct the masses in the technique of modern life; it has been found impossible to educate them. They have been given tools for an intenser form of existence, but no feeling for their great historic duties; they have been hurriedly inoculated with pride and power of modern instruments, but not with their spirit. Hence, they will have nothing

[8] Leviticus 18: 21, 20: 1-5.

to do with their spirit, and the new generations are getting ready to take over command of the world as if the world were a paradise without trace of former footsteps, without traditional and highly complex problems.[9]

Millennials are deficient in what I call, "critical participation." This means self-differentiation and empathy, without which ethics proves impossible. Self-recognition comes only from one's encounter with other individuals. Just as in the Holy Trinity each member recognizes the other: Father, Son and Holy Spirit in *One* absolute and universal essence, so in the human microcosm worth and value establish themselves through mutual reciprocity, which is the recognition of one's self as equal to but not greater than others—*diversity in unity*, "on earth as it is in heaven." It takes many people to form one person. Mankind, like the Creator is One family of persons—*plurality in singularity.* "Let *Us* make man in our own image, after our likeness . . . the two shall be one flesh."[10] Rabbi Martin Buber named this paradigm the *I/Thou* encounter, which subsequently became the foundational framework for 20th century Existentialism. Likewise, Hegel wrote that reciprocal self-recognition is the passage to, "Social Order" and ethics.[11] Unthinking and unfeeling technicians do not behave as responsible or mature citizens in the Modern Republic, nor do they make good ascetics in pursuit of enlightenment, nor lovers of wisdom in search of one honest person, neither can they function as individual members in the One Body of Christ. However, they make excellent soldiers.

Born on the cusp of global peace, the first computerized generation now find themselves at the precipice of world war, ecocide and human extinction. Generation Y, schooled by their predecessor X along with their protégé Z may literally, live out their name sake *X, Y and Z* in the coming decades by leading us into a technological abyss as they descend into a crucible of fire from which there will be, "no exit" — (Sartre). The transhuman future promises to be a world populated with human drones serving the hive mind. The *I/Thou* experience in which *one's dignity is recognized in the dignity of others* is not possible in a society of insects: one mind, one will, one world. Like under bees and worker ants people will be programmed genetically and connected *one-in-all* via electronic impulses and light vibrations beamed directly into the brain and central nervous system. *Humanity 2.0* will prove to be a sexless, mindless undifferentiated mass of flesh and carefully concealed electronics, prosthetics, hidden computer chips and invisible cameras.

An Old Russian proverb says, "The fish rots from the head down," signifying the logical order to social corruption. As our leaders go, so goes the country. Group pride is that solitary sickness endemic to the nation; it is

---

[9] Gasset, *The Revolt of the Masses*, 51.
[10] Genesis 1: 26, 2: 24; Matthew 19: 5.
[11] Hegel, *The Phenomenology of Mind*, V.B,1.

expressed by intolerance for difference. We know it best as that, "great Leviathan . . . which is but an Artificiall Man" — (Hobbes), whom I hereby rename, "Machinehead" for an accelerated generation. This is the source of war and injustice in our times, not a defect in individual character or a natural propensity to violence. Compounded in the crowd sin acts as one person without conscience and easily manipulated with fear and deception. "Madness in individuals is something rare, but in groups, parties, nations and epochs it is the rule." — (Nietzsche).

Historian Lewis Mumford demonstrated that the origins of war lay not with human nature as innately violent; a rationalization that declares war inherent to the nature of society and therefore, inevitable. We then blame an abstraction and evade moral accountability and limits, guaranteeing our self-destruction. War as we understand it today, nation against nation, begins with the *god-kings* (politicians) of ancient Mesopotamia and Egypt. Militaristic kings imposed a political structure on the *Social Animal.* They organized and exploited their human resources through concentration of *money*, *religion*, *power*, *slavery* and *technology* in a capital city (*polis*). Thus, creating a caste system to reinforce their absolute power and wage more effective warfare. The great cities of Babylon, Nineveh, Egypt, culminating in Rome were War Machines (Mega Machines) of the highest rank run by men who claimed to be gods on earth and had the fire power to back it up. "The melech [Moloch: King Nebuchadnezzar] spoke, and said, 'I AM BABYLON THE GREAT by the work of my hands and for the glory of my Name I built this palatial city.'"[12] "War," Mumford reminds us, "was an integral part of the constellation of civilized institutions held in tension with the city, on the basis of the division of classes, slavery and forced labor and religious uniformity. To remove any part of the fabric seemed to the rulers of men, a threat to every other part. They exalted the sacrifice of war. They wanted to maintain their own power."[13] War is organized insanity, the more organized we become the more proficient in killing.

In the Bible God rejects *political absolutism*: Oriental Despotism, Occidental Fascism, Monarchism and Imperialism as legitimate government because its warlords vie with Him for adoration.[14] The king's equivocation of

---

[12] Daniel 4: 30.

[13] Lewis Mumford, "How War Began," *Saturday Evening Post* April 18, 1959.

[14] First Samuel 8; Isaiah 14; Ezekiel 28; Psalms 82; Matthew 4; Luke 4; Revelation 17. First Samuel clearly outlines that God accommodated Israel's evil desire for a king because they had rejected God as their king. But even after the Monarchy was established the prophetic office remained in authority over the king. Israel's king was never considered absolute as with gentile kings. The people's morality over-rode royal decree and the prophet confronted the king with his sin (I Sam 14: 45, 46; 15: 23; 22: 17; 2 Sam 12). Likewise, the stories of Esther and Judith are examples of blatant political subversion. The Greeks, also, in the story of *Antigone* suggest that divine laws supersede human ones.

himself with the star-god was tantamount to idolatry and Luciferian worship. In reference to the king of Babylon Isaiah declared, "How are thou fallen from heaven, O Lucifer [day-star] son of the morning . . . 'Is this the man that made the earth to tremble that did shake kingdoms.'"[15] Anyone who fancies himself by mere conceit to be a god—master of the universe—places himself in direct opposition to the God-Man Jesus Christ, who called himself, "the bright Morning Star"[16] who will defeat the Luciferian forces of man worship with the power of his word, "I smote them through the prophets, I slew them by the words of my mouth," Hosea pronounced. "He shall strike the ruthless with the rod of his mouth and with the breath of his lips he shall slay the wicked," Isaiah exclaimed. And John the Revelator echoed, "his name is called The Word of God . . . out of his mouth goeth a sharp sword, that with it he should smite the nations: and he shall rule them with a rod of iron."[17]

Ancient Israel's embrace of monarchy was an affront to God. "They have rejected me that I should not reign over them."[18] Israel wanted a king for the expressed purpose of enforcing a strict social order to wage war. "We are determined to have a king over us, so that we also may be like other nations and that our king may govern us, and go out before us and fight our battles."[19] Fascism always arises in an age of decadence, through youth and hygiene movements filling a political and spiritual vacuum, with a preoccupation with bodily purity and national self-interest.[20] Like the undead risen from their shallow graves the dark forces of history: *religion, money, power, slavery* and *technology* are alive and well on planet earth. If allowed—*only he who now letteth will let*—free reign, they will stop at nothing less than destruction of everything. This self-same constellation of powers from antiquity has returned on a colossal scale in our *technosecular* world to wage war against God and His Christ through the murder of the image of God in humankind. Moloch demanded sacrifice of the first born to insure against the advent of the Christ child. Today the Machine-God will attempt to remake that child in its own ephemeral image. All that is missing from the contemporary scene is the appearance of the star-god, *the transman* of sin in the image of the beast.

---

[15] Isaiah 14: 12-16 (KJV).
[16] Revelation 22: 16 (NIV).
[17] Hosea 6: 5 (NAB); Isaiah 11: 4 (NAB); Revelation 19: 15 (KJV).
[18] First Samuel 8: 7 (KJV).
[19] First Samuel 8: 19, 20 (NRSV).
[20] Judges 21: 25; First Samuel 8: 1-18.

Introduction

This book is divided into three parts that represent my definition of technology. Part I, Imagination, technology begins with great ideas, fantastic visions and dreams, like voyages through space and instant communication, not very practical at first, but the seed of knowing has been planted, a knowledge that will flourish into a useful artifact. Through trial and error, it grows, taking on a life of its own. It makes itself indispensable. Like all slaves it forms a symbiotic relationship with its master. A relationship that does not remain static, but dynamic and as in traditional society the servant usurps his lord, so in simulation, technology rules its creator. We remain in smug complacency sure of our mastery over the world never dreaming our busy distraction will replace us. Chapter (1), discusses the origins of the space race in film and fantasy that fuels our current preoccupation for a faster tomorrow. (2), likewise, locates utopian metaphor in visions for the 21st century, which depicts the city of the future as mega structure ready to depart from Earth. Posthuman discontent with our current limitations was logically imbedded in the Modernist Project. Part II, Knowledge, knowledge is awareness of the continuum of history and our place in it; it is the unveiling of truth in our consciousness and is therefore never final but remains open to future possibilities. In chapter (3), we begin to see the shape of things to come take form in the rise of Machinehead. Machinehead is the personification of artificial intelligence in the near future. If I were to choose a fictional example, it would be D.F. Jones' two novels *Colossus* and *Fall of Colossus*. Colossus is the literary type to Machinehead, a supercomputer assumes control of human affairs; an example of the biblical image of the beast, which was not a real person, but the simulation of one in the form of an automaton. In the 21st century we call these political projections talking heads, which are media fabrications. They are also personalities we currently ascribed to Artificial Intelligence whenever we address our computers by name, distractions that take our attention away from their slow but steady encroachment into our lives. Machinehead is the ultimate talking-head, and all-knowing avatar. However, unlike the fictional example, we can't imagine that computer scientists and engineers will be as stupid as Father Forbin, Colossus designer, and neglect to put a fail-safe mechanism into AI. An off button for when things begin to slip from their control. But even then, an off button does little more than impart a false sense of security, a clever illusion used to fool us into thinking we retain mastery over a force we have suddenly realized we can no longer live without. Honestly, how many of us can afford to turn off our electronic devices, even for a couple of days? The more we use them the harder it will be to go without. AI works the same way as your computer, cell phone or automobile—use turns into habit and habit into need and need into addiction. Chapters (4), (5), and (6) offer the substance of my argument. *We are speedily deifying technological simulation that is replacing our traditional and natural environments.* What is ultimate means what is most important. Our view of God, nature, society, humanity and our place in history comes from the tradition we were raised in, such as church, family and friends, as well as our natural milieu: the farm, the countryside, suburbia or cityscape. What would your view of God be if raised on the Moon or Mars, perhaps, a harsh desert deity compounded with social isolation, emotional stunting which limits empathy by lack of familiar contact. This leads us to Part III, Spirituality, spirit is not matter, but stands above it and infuses it with life in a way we do not fully comprehend. Truth and therefore the lie are the two poles that act as the axis point in reality. One path leads us to God and the other to idolatry, which is self-glorification according to the lie that we can serve as our own god without self-destructing. Idolatry has always been the mainstay of the world's religious experience. The human heart is an idol factory, Calvin said. Machine worship is no different. In chapter (7), we see that Machinehead will lead to the same social void as the ancient imperial cult, which is ultimately man-worship. Chapter (8), there is a killer robot in the bible, conventionally, called, "the image of the beast" (Rev 13: 15). This is not the emperor, but his likeness in the form of an animated statue, which became his public persona and was venerated as such. This is the ultimate in Distance-Worship. The imperial cult could spread itself to all regions in the empire through public adoration centered on the similitude of God on Earth. As Colossus is the fictional parallel so the image of the beast remains the scriptural analogy to Machinehead. It looks like the emperor, it even talks and moves like a man, but it's not a man, it's a machine impersonating a man to fool the masses. (9), is my take on the next highly anticipated public event: Singularity and the critical role the Millennial generation will play. This is when humanity and machine become one essence, yet without their own distinctions—a false apotheosis. What we should expect from idol worship. The flesh is not raised to divinity as in Christ's resurrection, but will be downgraded to something obsolete. It is always wrong to put a date on any event in the future that claims ultimacy, such as AD 2045 because it distracts from present reality which is responsible for future things. I reject technological determinism inherent to Transhumanists expectation that believes the merging of mankind and machine to be inevitable. The future is only fixed if we fail to see that it's controlled by what we do now. 2045 will come and go with a whimper because the big bang of the end of the world has already happened with the splitting of the atom. This event in conjunction with AI and ICBM's remains the biggest danger. Our destiny is now subject to a technology that must never be used, yet, there is no such thing as useless technology. This puts nuclear power and its weaponization beyond the reach of *Just Cause*. Technology is now genocidal.

Chapter (10) is about the eternal now. The moment we realize we control our own destiny, not some impersonal force that directs all into its irresistible vortex. People determine the future not accelerated change. Let those who want to join the machine go. What is that to us? Space is their ultimate destiny, but it must not come at the expense of earth. The ultimate sin is to sin against the Earth Nietzsche preached. If space exploration cannot be self-sustaining it must be abolished. Commonsense tells us the vacuum of space is a death trap to all who challenge Earth's gravity. The present tense is the determining element not the future or the past. If we care for today, tomorrow will care for itself. This will happen if we adopt a critical perspective and stop foolishly assuming all acceleration is good. The freedom to disagree will be the decisive factor in the race to the future.

Preliminaries

Prologue

to

Human Extinction

If you could save the world from nuclear extinction by giving up your cell phone, even if it took 1,000 years, to end proliferation would you do it? If I told you that giving up the cell phone was worth the salvation of your soul, you would find a way to do it. The proposition before us is simple but is not easy. The consequences for success or failure will be sublime. Abolish the cell phone and obtain world peace for future generations or stupidly indulge the virtual pleasure of the flesh for immediate gratification, like mindless drones, sheep to the slaughter. We sacrifice Tomorrowland to the gods of war today. Abstaining from cell phones will serve as a protest against nuclear weapons; whereas we cannot as yet find the will to abolish nukes, we can, however, mitigate cell phone addiction. This is no false choice. Experts have been telling us for a decade that the cell phone is the gateway to the cyborg condition. This is the point of technological convergence, inevitably shrinking to the size of a mustard seed for hypodermic insertion. An electronic Pied Piper will lead the present generation into the Termitarium to slave for their Queen. There is no freedom in the Hive, no privacy, no individuality, zero tolerance for differences, no originality, no growth, totalitarian democracy. There is no *Us/ Them*, no *I/Thou*. There is only *We*. And *We* crave war. War is collective will of the people, the engine of progress, life blood of the nation. Sun Tzu said, "The art of war is of vital importance to the state." *War is politics by other means* according to Kommandant Carl von Clausewitz and *those means are technology*. "War is thus an act of force to compel our enemy to do our will." Engineer Samuel Florman noted, "Engineering, for almost all recorded history, was closely linked to the military. Fortifications and weapons were major engineering concerns. Transport and water supply came with the province of military planning. The term *civil engineer* did not even exist until the mid-eighteenth century."[21] Technology is war. Religious philosopher Nikolai Berdyaev argued that as we accelerate technologically the tendency will be to slip all ethical bonds in the escalation of war.

---

[21] Samuel C. Florman, *Blaming Technology: The Irrational Search for Scapegoats* (New York: St. Martin's Press, 1981), 45.

"Scientific discoveries and technical inventions represent the terrible danger of more and more war."[22]

A desperate struggle for survival between enemies defines reality; victory belongs to the party that exercises greater force and, "force . . . equips itself with inventions of art and science." AKA: technics, industry, technology, the mechanical action of people without spirit or principle to support and guide its acceleration that has no purpose outside itself. War feeds on itself like any automatic process until there is nothing left. "War is an act of force. Each side, therefore, compels its opponent to follow suit; a reciprocal action is started which must lead in theory to extremes."[23] Escalation in war is inevitable according to the dictates of the law of progress. The only way to stop war is to stop fighting, to deconstruct the means of war, reverse acceleration. The war to end war has brought us to the brink of self-destruction. War begets war, not peace. We cannot stop war, genocide and ultimately universal immolation by accelerating the means of war. We make war inevitable. Guns were made to shoot; computers were made to deliver Hydrogen bombs with pinpoint accuracy. Technology's application to civil society underscores the materialist outlook that believes all people are naturally at war with each other, "the war of all against all."[24] Therefore, we are in need of an absolute dictator to provide order and security: the modern Moloch, Leviathan, *Machinehead.*

Technological acceleration has outpaced our collective will to resist it. At present rate *We the People* will end in a mushroom cloud. It's just easier that way, as the French say, "*après moi le deluge*" (after me the flood). We will consume and party to the fullest today and to hell with the future, "eat and drink today for tomorrow we die!" We know what the right thing to do is. We just don't want to do it. The hour is late. The time to decide is here. We have everything to lose and only the hell of nothingness to gain in a future ruled by Machinehead. I challenge you awake and rise up! Throw away your invisible chains and turn away from the abyss, launch out on the enterprise of self-discovery, explore the undiscovered regions of inner space, without the mediation of a digital illusion dictating what image you must conform to, you will surprise yourself.

---

[22] Nikolas Berdyaev, *The Realm of Spirit and the Realm of Caesar* (New York: Harper, 1952), 22.

[23] Carl Von Clausewitz, *On War* 1.2 (1832).

[24] Thomas Hobbes, *Leviathan*, 1651.

Blaming Technology

I tell you one time, you're to blame.[25]

We cannot condemn the atrocities of war, genocide, ecocide, or the propaganda machine—media that excuses this behavior through delusional imagery, faulty logic and partial truth—without also indicting their instruments as causal agent. The technological agents make these practices possible and well-nigh inevitable. Norman Cousins the famous social commenter, witnessing the advent of the atom bomb reminded us, "if history teaches us anything it is that the possibility of war increases in direct proportion to the effectiveness of the instruments of war."[26] If we wish to stop certain ends, such as extinction and genocide, we must curtail their means. Limit the means to affect unjust ends. Without the technology the ends do not follow. Means and ends are one. "The means to the end are not unlimited, for the end itself sets the limits in each case," Aristotle said.[27] In his monumental tome, *War and Human Progress* Professor John Nef argued that war has become deadlier since the industrial revolution given the increasing speed of technological acceleration.[28] Speed kills! Literally! Jonathan Schell wrote, "In the centuries of the modern scientific revolution, scientific knowledge has steadily increased the destructiveness of warfare, for it is in the very nature of knowledge, apparently, to increase our might rather than to diminish it."[29] Professor Hugo de Garis demonstrated in his book *The Artilect War* that with each passing global conflict from the Napoleonic wars to World War Two the body count has multiplied exponentially from hundreds of thousands to tens of millions and projects that the next world war will yield a horrific death toll in the billions measured as *gigadeath*.[30] Sigmund Freud spoke prophetically of a real cultural lag between our technological prowess and lack of spiritual restraint that has resulted in two world wars and threatens survival with a third, "Men have gained control over the forces of nature to such an extent that with their help they would have no difficulty in exterminating one another to the last man."[31] Einstein said of nuclear weapons that, "the unleashed power of the atom has changed everything save our modes of thinking, and we thus drift towards unparalleled catastrophes."[32]

---

[25] Rolling Stones, "Sympathy for the Devil" *Beggars Banquet* (Olympic, 1968).

[26] Norman Cousins, *Modern Man Is Obsolete* (New York: Viking, 1945), 42.

[27] Aristotle quoted in Jonathan Schell, *The Fate of the Earth* (New York: Avon, 1982), 189.

[28] John U. Nef, *War and Human Progress: An Essay on the Rise of Industrial Civilization* (New York: Norton, 1968).

[29] Schell, *The Fate of the Earth*, 106.

[30] Hugo de Garis, *The Artilect War: Cosmists vs. Terrans: A Bitter Controversy Concerning Whether Humanity Should Build Godlike Massively Intelligent Machines* (Palm Springs, ETC., 1995).

[31] Sigmund Freud quoted in Schell, *The Fate of the Earth,* 155, 156.

[32] Albert Einstein quoted in Schell, *The Fate of the Earth,* 188.

Philosopher Gabriel Marcel insightfully connected war, technical knowledge and mass suicide. "War with the face it wears today is sin itself . . . war is becoming more an affair of technicians . . . everything that gives a new impetus to technical research at the same time renders war more radically destructive, and bends it more inexorably to what, at the breaking point, would be quite simply the suicide of the human race."[33] Nuclear war is a high-tech form of cannibalism. Disarmament, then, is the only effective Christian political position. Every other view, such as Socialism, Capitalism, even Democracy does not need Christianity to function. They all operate according to political realism founded on the law of fear and war over against the law of love and peace.

Technology is not a neutral gift that descends on us like manna from heaven; rather technology is a moral contrivance consisting of imagination-knowledge-spirit, ambivalent by design, a tool, really a system of tools and as every god-fearing gun owner knows, every tool, doubles as a weapon. This means judging technology as efficient cause from its good and bad effects will always yield morally ambiguous results; for example, the industrial revolution serves as the agency for both increased longevity in times of peace and a higher body count in time of war, as well as, threatening human extinction as one potential future and promising global survival as the other. As extension of human ingenuity, brilliance and pride technology can no more be neutral than your hand or foot or eye. "Technology is an expression of the aggregate communal will," Florman said.[34] We only have ourselves to blame when our bombs kill women and children, not the enemy and supergerms evolve resistance to our most advanced medicine. *Technology is Us accelerating faster and faster into the future.*

What Is Neutral?

I turn wise men back and make their knowledge foolish.
Isaiah 44: 25
(NAB)

The basic mistake in the popular conception is expressed as, "technology is neutral; it depends on how you use it." The majority who pause long enough to think about the consequences of their use simply repeat this mantra with the greatest of ease and then put the issue to bed. However, they fail to realize that they just swallowed a glow-worm as bait, hooking them to the treadmill of acceleration. The neutrality argument explains nothing and merely serves to excuse us from self-examination by inducing acute brain paralysis coupled with untoward flatulence from the mouth. Its cousin in thought, the argument from necessity, fails to do justice to our practice since,

---

[33] Gabriel Marcel, *Man Against Mass Society* trans. By G. S Fraser (Chicago: Gateway, 1962), 80, 81.

[34] Florman, *Blaming Technology*, 53.

"necessity" is not an argument because it does not give place to free choice; it is however, the lowest form of ethical defense. Neutrality is akin to technological determinism that asserts, "progress is inevitable," which also denies freedom. Thus, we are condemned to the illogical conclusion that we have control over a neutral technology moving in an inevitable direction.

"Neutral" is a word that has no correspondence to anything outside itself. *The Oxford English Dictionary* defined, "neutral" as, "having no strongly marked characteristics or features, undefined, indefinite;"[35] likewise, according to *Webster's*, "neutral" means, "not taking part."[36] *Neutral* is a word that explains nothing about the prime determinant in contemporary history: *technological acceleration*. William James, the American Philosopher explained, "neutral" as describing what we can never know. "Matter we know and thought we know and conscious content we know, but neutral and simple 'pure experience' we know not at all."[37] Professor Hegel neatly put it, "a sign indifferent to what is signified . . . in reality [signifies] nothing."[38] "Neutral," then, is less than helpful as a description of technology and serves as a ruse to get past critical self-examination. If anything, "neutral" describes what technology is not. *Technology is not neutral.*

A tree is known by its fruit, Jesus said; a good tree will produce good fruit, conversely, a bad tree will produce bad fruit and out of the heart comes good and evil.[39] And, "from the wicked comes forth wickedness" David intoned.[40] Corruption is endemic to technology, just as it is to human nature. The Apostle declared, "There is none righteous, no not one . . . there is none that does good."[41] Accomplishments great and small are not exempt from this grim diagnosis. Technology is only as innocent as the people who design it, make it and use it. There are no good or bad technologies, any more than there are good or bad people. Jesus protested the rich politician who flattered him with the accolade, "good teacher" assuring us, there is no one good—except God.[42] Theologian Reinhold Niebuhr stated that the doctrine of human sinfulness was verifiable by empirical evidence.[43] Bad people cannot make good technology. And, "good people" cannot make bad technology

---

[35] *The Oxford English Dictionary*, 2nd Ed., Vol X Moul-Ouvm (New York: Oxford University Press, 1989), 356-57.

[36] *Random House Webster's Unabridged Dictionary*, 2nd Ed., (New York: Random House, 2001), 1292.

[37] William James quoted in *The Oxford English Dictionary*, 356-57.

[38] G. W. F. Hegel, *The Phenomenology of Mind*, trans., by J.B. Baillie (New York: Harper: 1967 [1807]), 345-46.

[39] Matthew 7: 15-20; Luke 6: 43-45.

[40] First Samuel 24: 14 (NAB).

[41] Romans 3: 10, 12 (NKJV).

[42] Matthew 19: 16, 17.

[43] Reinhold Niebuhr, *Moral Man in Immoral Society: A Study in Ethics and Politics* (New York: Charles Scribner's Sons 1932).

better. "Wisdom is vindicated by her works," Jesus said.[44] We can only judge technology on the basis of its results, not a presumed paradigm of inherent neutrality. What has technology actually produced on a global scale? What type of world has it created? What have been the undeniable consequences of technological acceleration?

The Hypnotized Never Lie

With their tongues they have practiced deceit
Romans 3: 13
(NKJV)

The biggest lie is the one we tell ourselves. It is a commonplace evasion today to divide the results of consumption into good or bad categories, as if use was immune to side-effects. We feel righteous in our use because we believe our cause is just and our motives pure when we put technology to, "good use." People will sooner blame *human error* in an orgy of misanthropic self-deprecation, rather than admit there may be a *design flaw* that necessitates limited use to mitigate negative side effects of their favorite devices. Automobiles are a blight on nature. Cell phones emit radiation. TV limits intelligence. Pollution is inherent to the nature of technology. To use is to pollute. We lie to ourselves rather than recognize the inseparability of good and bad effects. The, "right use" will produce the same results as the, "wrong use." Users are hypnotized by their addiction.

What novelist Aldous Huxley called, "hypnopedia" the infinite repetition of slogans, mantras and catchy phrases that encapsulates a truncated version of truth proliferates through media saturation. This highly effective form of propaganda works in tandem with each person as a form of hypnosis. Propaganda is hypnotism. But hypnotism doesn't work if you know you are being hypnotized. Huxley demonstrated, "words without reason . . . the kind of words that require no analysis for their comprehension but can be swallowed whole by the sleeping brain . . . used for moral training . . . for conditioning of behavior through verbal suggestion at a time of lowered psychological resistance."[45] When we are tired and worn out from a busy day at work and relax in front of the TV, we allow a materialist gospel to subconsciously program us. Commercials are not separate entities that stand alone from regular scheduled programming, but an integral element to the conditioning process. Subtlety is the operative principle in mind control. Pity jingles bolstered by sex appeal and high drama conceal an otherwise indigestible message; sugar coated to go down pleasurably. This is similar to master political satirist George Orwell's sloganeering in his dystopian novel *1984*:

---

44 Matthew 11: 19 (NAB).

45 Aldous Huxley, *Brave New World Revisited* (New York: Perennial, 1965), 70.

WAR IS PEACE

FREEDOM IS SLAVERY

IGNORANCE IS STREGTH[46]

These counter-intuitive psycho-tropes give the appearance of commonplace wisdom but are vicious mind-traps that paralyze reflective self-criticism and serve to excuse our participation in an accelerated social movement that appears hopelessly out of control. Ours is not unlike the world of Oceania:

YOU CAN'T STOP PROGRESS

GO WITH THE FLOW

GREED IS GOOD

The radical dichotomy that posits a sharp division between, "bad people" and, "good technology" i.e., "neutral technology" blinds the hearts and minds of users to the subterranean spiritual powers not readily discernable to common-sense realism, but determinative, nonetheless.

New World Man

We are misquoted as saying, 'Let us do evil that good may come.'
Those people will suffer God's righteous indignation.
Romans 3:8

A generation ago, *New World Man* ascended from millennia of barbarism as the indomitable Nordic hero, *the Last Man*, chained as it were, by iron manacles in a steel cage of his own forging. Today, as consequence of *the Great Instauration* (restoration of ancient knowledge) promised by Sir Francis Bacon, famous for his well-worn quip: KNOWLEDGE IS POWER, *New World Man* stands at the ready to turn his earth shattering, heaven shaking technology on himself. Like *Oedipus Rex* of Greek Tragedy fame mutilated himself as punishment for tempting fate and as proud king Saul, fell on his own sword, so *New World Man* fresh from victory over nature and liberation from tradition has embraced suicide as the inevitable tragic end to his mechanical philosophy. We have surrendered freedom and inherent human dignity, for the nebulous goal of finite conquest of an infinite space desert. An empty region we can only look at but will never exploit. Just as Esau traded his birthright for warm soup and Doctor Faustus sold his soul to the devil for knowledge and power, *New World Man* has abdicated his right to exist; in exchange for freedom we are rewarded with momentary security, crass material comfort and ease. Our soul grooved by visions of absolute victory or total defeat. Imagination is imprisoned by the spiritual powers of media, so limited are we in thought, so captive in our imagination to the prevailing

[46] George Orwell, *1984* (New York: HBC, 1949), 3, 14, 23.

cultural motif of technological progress that we can scarcely remember or conveniently forgotten, a future in millennial peace under the righteous tutelage of Our Lord Jesus Christ, King of Kings. His reign does not encompass technological superiority but embraces meekness. The prophet exclaimed, "your king is approaching, he is vindicated and victorious, humble and riding on a donkey . . . He will banish chariots from Ephraim and horses from Jerusalem; the bow of war will be banished. He will proclaim peace to the nations, his empire will stretch from sea to sea."[47] Technological mindsets are not fixated on peace, but war. Our minds shackled to a future world vision where God is dead, humankind is extinct, and the machine inherits a dead planet.

The advent of the atom bomb punctuated history brining mass confusion between means and ends, cause and effect and marking the final era in a long historical movement of pride and man worship. The ultimate fruition of *Homo Deus* (man making himself God),[48] which represents a world turned upside down and inside out, where all traditional values are reversed, ends and means are inverted, and effects create their own cause. The invention of the atom bomb crossed the threshold of spiritual reversal. Scientists wanted to taste the, "technically . . . sweet" fruit of the tree of knowledge as Oppenheimer put it.[49] Engineers took a risk for the sake of knowledge under political cover that was unnecessary and opened the portals of the dark abyss unto their sons and daughters.

Philosophers of Technology call this transvaluation—equivocation of good and evil: The Technological Imperative—what can be done must be done. Acceleration exists for its own sake and humanity is downgraded to merely the agent for further change, like a cog in a wheel. Because the ability to do something exists, such as fly to the moon, spilt the atom, splice the gene, upgrade electronics, we feel compelled to do so, as an unthinking reflex—like moth to flame. Genocide becomes acceptable if it serves a greater cause. We have lost central place in the universe demoted to object of science and technology rather than subject. The 2,460 plus warheads let loose since Trinity on land, in the sea, under the earth and in the sky were petty ante, incomparable to the price our children will be forced to pay for a risk in the cause of science made by the *Greatest Generation*, who will be remembered as the *Infamous Generation* who murdered the future to save themselves.

---

[47] Zechariah 9: 9, 10 (NJB).

[48] Yuval Noah Harai, *Homo Deus: A Brief History of Tomorrow* (New York: Harper, 2017).

[49] J. R. Oppenheimer—Father of the Atom Bomb—quoted in Gerard J. De Groot, *The Bomb: A Life* (Cambridge, MA: Harvard University Press, 2005), 176.

Society is the human tool in the service of technology; for what? We know not. Ours is not to wonder why, but to do. We now assume the position of servant to LORD TECHNOLOGY.

Welcome to Machinehead

There is no fear of God before their eyes.
Romans 3: 18
(NKJV)

Victorian satirist Samuel Butler foresaw a darker potential future than either Orwell or Huxley imagined. A time will come, Butler screamed when the work of our hands will turn against us, a future in which *the natural order of things* will be inverted, turned upside down, which is theologically speaking, the hallmark of true Satanism, yes there is a false Satanism. In his essay, *Darwin Among the Machines* (1863) Butler introduced the world, in the words of MIT History Professor Bruce Mazlish, "By an extraordinary leap of intuition . . . [he] dramatically announced the theme of the machine as an advanced evolutionary species."[50] By means of what today we call, "Artificial Selection," machine, "Breeding" Butler believed will produce an artificial intelligence beyond our control. The machine will eventually attain autonomous status, "by all sorts of contrivances that self-regulating, self-acting power which will be to them what intellect has been to the human race."[51] Before Darwin's theory of Natural Selection could solidify in the gizzards of Tennessee Daytonists and Darwinists alike, Butler sounded forth the alarm that at the end of the Victorian love affair with the machine, "We shall find ourselves the inferior race."[52] We will be to our machines as domesticated animals are to us today. We feed and care for our pets. They have cultivated a symbiotic relationship with us. We do the same with our machines. Professor Mazlish described, "We will have to feed them and in doing so become unto them as a kind of 'Slave.' Thus, just as we care for animals the machine will care for us; after all, we are useful to them. Put simply, machines will have dominion over man . . . we are creating the means of our own inferiority and enslavement"[53] ; that is, *by our own choices.* Thus, the machines are not taking over in some sort of *robo coup d'état*, which is ridiculous. *I repeat the machines are not taking over.* However, weary of freedom, life, love, thinking, decision-making and self-determination, *Sapere aude*—no more, in the absence of hope, we joyfully resign ourselves to their tender mercies.

---

[50] Bruce Mazlish, "Butler's Brainstorm" in *Prefiguring Cyberculture: An Intellectual History*, Ed. Darren Tofts, *et al.* (Cambridge, MA: MIT Press, 2003), 230.

[51] Samuel Butler quoted in Mazlish, "Butler's Brainstorm" 230.

[52] Ibid.

[53] Mazlish, "Butler's Brainstorm" 230.

"Aghast at this vision, Butler has his presumed author, Cellarius, exclaim, 'war to the death should be instantly proclaimed against every machine.'"[54]

Sic Semper Tyrannis

DEATH TO MACHINEHEAD!

---

[54] Samuel Butler quoted in Mazlish, "Butler's Brainstorm" 230.

Part I

Imagination

The future of the past is in the future.

CHAPTER ONE Space the Final Frontier

Imagination Anticipates the Unknown

Tomorrowland begins today. The world we want to leave for those who follow us starts right now; with every imaginable thought, we plant a choice, like a seed that will flourish into action and with every act, we sow our legacy, which is a destiny the future will harvest. We have the technology to do nearly anything we desire, but not the-will-to-do-good. We would sooner kill for our cause, then, die to ourselves. We destroy the future to live comfortable today. We will not inherit the Earth. Kill or be killed leads to cannibalism—war of each against the other is not God's will for us. The-will-to-do-good is through the-will-to-believe in him who said, "Do to others as you would have others do to you." Christianity is the true humanism because we declare the human-angelic future on earth as it is in heaven; not a desolation in outer space. The meek do not kill their enemies rather they convert opponents into powerful allies, making themselves twice as strong. The mighty on the other hand, murder their foes. They lose the opportunity to make friends, spending their strength, making themselves twice as weak. We must follow the Golden Rule and not the Law of the Jungle, if we wish to be called children of the Heavenly Father.

What Type of Future Do We Want?

The Red-Hot Chili Peppers sing, "Space may be the final frontier, but it's made in a Hollywood basement."[55] These lyrics suggest that the contemporary dream of infinite space exploration is a perceptual social construction programmed into our imagination *via* media magic. The latest cinematic *futuribles* (studies in conjecture) downloaded into popular consciousness prepares us for the advent of the Hive Mind, AKA Cyborg, *vis-à-vis* the Singularity (*circa* AD 2045). Films such as, *Transcendence*, (2014) *Blade Runner*, (2017) *et al.*, reflect the present generation's imagined future,

---

[55] The Red-Hot Chili Peppers, "Californication," *Californication* (Warner Bros., 2000). American Historian Frederick J. Turner, defined *Frontier* as, "the meeting point between savagery and civilization." *The Significance of the Frontier in American History* (1894); https://www.historians.org/about-aha-and-membership/aha-history-and-archives/archives/the-significance-of-the-frontier-in-american-history

its fears and hopes in the computer age, collective dreams and nightmares projected back to the future of the latter half of the 21st century. Tech-Noir envisions the future as the end of human history, its zenith, complete and perfect in the Singularity, which will look something like total carbon freezing. Imagine Earth reduced to a giant Shrinky-Dink, *à la Men in Black*, (1997) sealed in silicon, unable to move or grow, a plastic ornament to hang off Orion's Belt. Like ants discovered in Baltic amber fossilized sixty million years ago, they remain unchanged, forever the same. History will be frozen according to the law of progress, which says, acceleration must reach a high point and then plateau before regressing. There is no such animal as unlimited technological returns. Acceleration has a stopping point, the faster we move the closer we get to it. Beyond that point it's impossible to see.

Professor I. F. Clarke related in *The Pattern of Expectation* that today's obsession for future change abruptly started 200 years ago when the Industrial Revolution introduced an imaginative new awareness that the future will be better than the present by reason of technological innovation and progress that, "transformed the condition of human life throughout the planet. The everyday experiences of social change and industrial development has become so much part of our general thinking that we are apt to forget the extraordinary novelty of seeking to discern the shape of coming things. The frequent projections of tomorrow's world in films and television programs hide the fact that for most of human history the image of the future has been a blank."[56] Technological progress has allowed contemporary times to choose from an array of potential futures. Futurist scholar John McHale explained that following the Industrial Revolution, there was "a growing realization that man's future may be literally what he chooses to make it, and that the range of choice and degree of conscious control which he may exercise in determining his future are unprecedented."[57] Professor of Government and distinguished Futurist Victor Ferkiss stated, "humanity is on the threshold of self-transfiguration, of attaining new powers over itself and its environment that can alter its nature fundamentally as walking upright or the use of tools . . . [through] modern technology . . . giving man almost infinite power to change his world and himself."[58] Leading Futurist Bertrand de Jouvenel asserted that a major focus essential to human mastery over technology concerns what type of future that society wants from its technological progress; having its choice of futures, then, we must take the initiative to create what we want, rather than resigning to inevitable fate. "Finding out what we want should become a major object of attention . . . [T]here is a

[56] I. F. Clarke, *The Pattern of Expectation: 1644-2001* (New York: Basic Books, 1979), 2.

[57] McHale, *The Future of the Future*, 7.

[58] Victor Ferkiss, *Technological Man: The Myth and the Reality* (New York: Mentor, 1969), 28, 30.

vast difference between letting changes occur under the impact of technological advances and choosing the changes we want to bring about by our technological means."[59] Emmanuel Mesthene, former Director of the Harvard Program in Science and Technology, affirmed similar sentiments. "We have now, or know how to acquire, the technical ability to do very nearly anything we want."[60] Technology enables us to stagger population growth, reduce pollution, stop ecological decay, find alternative energy sources, cure poverty and disease, eradicate inequality and end war, "and make the world a far better place in which to live," and yet according professor Ferkiss, "we do not do so."[61] Popular belief asserts that technology can solve most of the world's problems. In the 1960's President Kennedy committed to win the space race and cure poverty at the same time. The first promise was easy to fulfill, the world is still waiting for the second.

Machine the Measure of All Things

Spin forever down the ringing groves of change.
Tennyson

The central tension in technosecular society may be traced back to the Italian Renaissance when, as McHale summarized, "the central argument was concentrated on man's relation to, and conscious control of, his own forward development."[62] The Renaissance turned to antiquity and nature for inspiration. However, humanity has lost center stage with the rise in Post-humanism, "soon after the Industrial Revolution, the summation of various scientific discoveries and access to large scale technical facilities tended to invalidate one side of the debate."[63] The rational, material and technological control over nature began to outweigh the spiritual, irrational (instinctual or intuitive), and religious aspects of Western heritage, "with the material means for physical survival and larger maintenance of human society augmented beyond any previous expectation, the possibilities and potentialities for a more conscious forward development became available."[64] Destiny, progress or perfection became associated with material excess and greed, envisioned exclusively in terms of the materialist, temporal and technological dimension. The traditional sacred was expropriated for the secular vision of technological utopia and totalitarianism as religious substitute. We slowly became the object and tool of science and technology rather than subject and master. The

---

59 Bertrand de Jouvenel quoted in McHale, *The Future of the Future,* 9.

60 Emmanuel Mesthene quoted in Ferkiss, *Technological Man*, 30.

61 Ferkiss, *Technological Man,* 35.

62 McHale, *The Future of the Future,* 6; Ernst Cassirer, *et al.*, eds., *The Renaissance Philosophy of Man* (Chicago: The University of Chicago Press, 1948).

63 McHale, *The Future of the Future,* 6

64 Ibid.

classical dictum: *man is the measure of all things* gave way to the Posthuman maxim: the *machine is the measure of man.*[65]

Space mythology, such as exploration, discovery and colonization serve the technological order as *metanarrative* or *myth* (Big Picture) providing society with a grand purpose and future hope. Allan Watts defined, "myth" as a word that does not mean something that is, "'untrue' or 'unhistorical.' Myth is to be defined as a complex of stories—some no doubt fact, and some fantasy—which, for various reasons, human beings regard as demonstrations of the inner meaning of the universe and of human life. Myth is quite different from philosophy in the sense of abstract concepts, for the form of the myth is always concrete—consisting of vivid, sensually intelligible, narratives, images, rites, ceremonies, and symbols. A great deal of myth may be based on historical events, but not all such events acquire the mythic character." [66] And Ferkiss affirmed, "the yearning for social purpose is as basic as are the related desires which cluster around the institutions of religion, and just as persistent. Social systems cannot long endure without a sense of purpose any more than can individuals . . . Human societies are self-conscious and must have a guiding vision of the future."[67] Philosopher Jose Ortega y Gasset exclaimed, "An 'unemployed' existence is a worse negation of life than death itself. Because to live means to have something definite to do—a mission to fulfill—and in the measure we avoid setting our life to something, we make it empty . . . Human life, by its very nature, has to be dedicated to something, an enterprise glorious or humble, a destiny illustrious or trivial . . . to live is to be directed towards something, to progress towards a goal . . . it is the something to which I put my life and which consequently is outside it, beyond it."[68] Berdyeav maintained that, "forms of authority . . . are always objectivized and rationalized sub-conscious states and passions. And this always means the creation of myths, without which it is impossible to govern human masses."[69]

---

[65] Erich Kahler, *Man the Measure: A New Approach to History* (New York: Meridian, 1943); Michael Adas, *Machines as the Measure of Men: Science, Technology, and Ideologies of Western Dominance* (Ithaca, NY: Cornell University Press, 1989); Merritt Roe Smith and Leo Marx, eds., *Does Technology Drive History? The Dilemma of Technological Determinism* (Cambridge, MA: The MIT Press, 1996).

[66] *Alan W. Watts, Myth and Ritual in Christianity (Boston: Beacon Press, 1968), 7; similarly,* *Metanarrative* means, "An overarching account or interpretation of events and circumstances that provides a pattern or structure for people's beliefs and gives meaning to their experiences: *'traditional religions provide stories that deliver a metanarrative about how we should live our lives';* https://en.oxforddictionaries.com/definition/metanarrative

[67] Victor Ferkiss, *The Future of Technological Civilization* (New York: George Braziller, 1974), 200, 201.

[68] Jose Ortega y Gasset, *The Revolt of the Masses* (New York: Norton, 1932), 136, 141, 142.

[69] Nikolai Berdyeav, *The Realm of Spirit and the Realm of Caesar* (New York: Harper, 1953), 81, 82.

Myth transforms abstract ideas into realities. We live by a collective consciousness that directs history to a particular end. A societal purpose or goal is tantamount to its *mission*, *destiny* or *imperative*. Imperatives are indicative of empires. Historical examples include *Pax Romana* (27 BC-AD 180) or the Roman Peace, the last great social vision of the classical world; an age eighteenth century scholar Edward Gibbon judged the, "most happy and prosperous" in human history.[70] The Romans believed the gods had bequeathed to them an empire without end in time and space: *imperium sine fine*. Present day examples are the *Pax Britannica* (1815-1914) the world under British control, history's first global Empire, and *mission civilisatrice* (1890-1945) for France; more popularly known as, "the white man's burden [to civilize the planet]." After *Manifest Destiny* the United States of America recognizes itself as the leader and defender of the free world, who's global *Democratic Imperative*, was captured by the slogan, "making the world safe for democracy." Currently, the USA leads the world in its own *Pax Americana* AKA *New World Order* (1945-Present). Subsequently, a *Pax Technologica* rules contemporary global society, which feels itself compelled by a *technological imperative*—what can be done should be done. Future global progress fixes technological acceleration irrevocably to space colonization as its *raison d'être*, reason for being, rather than the regeneration of planet Earth; for example, Krafft Ehricke the former Nazi and later American rocket scientist, astro-projected racial theory into cosmic consciousness by formulating a variant of the technological imperative that reaches beyond earthly limits. He called it *the Extraterrestrial Imperative* by which he argued for space exploration as necessary for human survival, a goal he hoped would provide the stimulus for limitless technological advance. "Ehricke's philosophy of the Extraterrestrial Imperative was based in part on the evolutionary tendency of living things to expand into new territories and in part on the need to transcend the limits of one small planet and assure the long-term future of humanity by utilizing the vast resources of space. He also hoped and predicted that the movement of humanity into space would usher in a new Age of Reason."[71] Ehricke formulated *Three Laws of Astronautics*, which enshrined into natural law our *limited* technological ability in pursuit of *limitless* space. "First Law: Nobody and nothing under natural laws of this universe impose any limitations on man except man himself. Second Law: Not only the Earth but the entire solar system, and as much of the universe as he can reach under the laws of nature, are man's rightful field of activity. Third Law: By expanding through the universe, man fulfills his destiny as an element

---

70 Edward Gibbon quoted in Kyle Harper, *The Fate of Rome: Climate, Disease and the End of an Empire* (Princeton, NJ: Princeton University Press, 2017), 24.

71 http://www.nss.org/resources/books/non_fiction/NF_079_extraterrestrialimperative.html

of life, endowed with the power of reason and the wisdom of the moral law within himself."[72]

The mystery of space is both the pull and the push into the unknown future. Professor Fred Polak discussed the dynamics of social change as a "push-pull process in which a society is at once pulled forward by its own magnetic images of an idealized future and pushed from behind by its realized past," he said.[73] Present reality and a hoped for or imagined future represent a dual nature whose perpetual dialectical struggle to balance opposites moves history and provides the momentum for social change. "Man's dualism is thus the indispensable prerequisite to the movements of events in time, and to the dynamics of historical change."[74] By, "dualism" we mean the, "unity of opposites,"[75] as in classical philosophy; for example, permanence and change, necessity and freedom, time and eternity, matter and spirit, etc.; in theology: Grace and Nature paradigm according to Saint Thomas Aquinas; in ethics the *is* and the *ought*. The visible, tangible and temporal world serves as manifest symbol or connecting point for the invisible, spiritual and eternal. Symbolizing activity distinguishes the human soul from the rest of nature. Metaphysics is the hallmark of humanity. Through maintaining balance and tension between complimentary opposites, or correlatives, one cannot exist without the other, the left and the right, the subjective and the objective, the past and the present, the future remains open: free, spontaneous, serendipitous, unpredictable; most importantly: *NEW!* The future is never fixed or certain, always becoming and subject to change, according to the human and divine cooperative will. "Thy will be done on earth as it is in heaven."[76] Balance is lost to uncontrollable extremes, as in our technosecular world, when only half the complimentary struggle is emphasized. Synthesis transforms into raging monism, the refusal to admit diversity while embracing totalitarianism that will inevitability return us to a pre-Christian view of man and society (All is One). The finite, the material, reason, money and technology usurp their conventional better halves: the infinite, the ideal, faith, prayer and the sacraments; complimentary opposites regress into, "irreconcilable opposites."[77] Technical reason becomes the exclusive organ for knowledge vilifying faith. Matter exorcises spirit degrading the temple of the body and

---

[72] Krafft Ehricke quoted from Gerard J. Degroot, *Dark Side of the Moon: The Magnificent Madness of the American Lunar Quest* (New York: New York University Press, 2006), 3.

[73] Fred Polak, *The Image of the Future*, trans., by Elise Boulding (San Francisco: Jossey-Bass Inc., 1973), 1.

[74] Ibid.

[75] Ibid.

[76] Matthew 6: 10 (KJV); the Apostle added, "we are God's co-workers" First Corinthians 3: 9 (NAB).

[77] Polak, *The Image of the Future*, 1.

denying the existence of the soul. Secular replaces sacred and technological acceleration devours natural resources.

Philosophical Materialism and its ideological subsidiaries: Atheism (God), Scientific Naturalism (Nature), Darwinism (Man), Marxism (Society) and Determinism (History) represent the collective world outlook in counter point to traditional Idealism (dualism) with its belief in two worlds, the ideal and the material. Marxist philosopher Maurice Cornforth put it this way, "Materialism . . . knows one world only, the material world, and refuses to invent a second, imaginary, superior ideal world. Materialism and idealism are irreconcilably opposed."[78] However, Princetonian Charles Hodge explained Materialism as a predominately-recent, "philosophical theory" that joined fatalism and atheism. "According to this system matter and motion are eternal; thought is an agitation of the nerves; the soul the result of our corporeal organization; the will the strongest sensation; the ground of morals a regard to our own happiness. There is no freedom, no morality, no future existence, no God. When these principles got hold of the popular mind, then, came the end."[79] Materialism accepts only one-half the traditional dialectic, creating what Russian religious philosopher Nikolai Berdyaev called, "ontological totalitarianism,"[80] or as Existential philosopher Gabriel Marcel said, "intellectual imperialism."[81] Reason ousts faith and matter rules over spirit. Berdyaev astutely observed that the origins of slavery are rooted in philosophical monism, only one element exists. "The philosophical source from which the slavery of man derives is monism. The practical expression of monism is tyrannical. Personalism is most profoundly opposed to monism. Monism is the domination of the 'common,' of the abstract universal, and the denial of personality and freedom."[82] Spiritual despotism rules over our technosecular world. This is an age in which spiritual life and traditional values have degenerated into crass materialism threatening the foundation of civilization with enslavement and ultimate collapse. Rather than, the amelioration of spiritual values along with material conditions, as legacy of technological progress, a relentless social determinism captures mass society in an irrevocable death grip represented by the modernist project's equivocation of mankind with nature. Reality's identification with matter

[78] Maurice Cornforth, *Materialism and the Dialectical Method*, 4th Ed. (New York: International Publishers, 1972), 23.

[79] Charles Hodge, *Systematic Theology*, Volume One (Grand Rapids: Eerdmans, 1985 [1871]), 247, 254.

[80] Nikolai Berdyaev, *Slavery and Freedom* trans., by R. M. French (New York: Charles Scribner's Sons, 1944), 51.

[81] Gabriel Marcel, *Man Against Mass Society* (South Bend, ID: Gateway, 1952), 155. Both Marcel and Berdyaev reflect the same macro-concept as *the Ontology of Technology* notion argued for by German Ontologist Martin Heidegger and French Sociologist Jacques Ellul's idea of *Automatism* as the defining characteristic of technological acceleration.

[82] Berdyaev, *Slavery and Freedom*, 68.

dehumanizes us by loss of human distinction from the material world. Homo sapiens become one creature among many. There is nothing special about us. An apparent inevitability moves all under its spell to peaceful resignation. This is exemplified by the commonplace metaphysical redundancy: It-Is-What-It-Is. Change then becomes symptomatic of a downward spiral, such as terminal illness running its course, or diminishing returns from progressive investment.

Into the Void

Rocket engines burning fuel so fast
Up into the night sky they blast . . .
Up into the black sky so vast.
Burning metal through the atmosphere
Earth remains in worry, hate and fear.[83]

Space mythology represents a teleological view that expresses human destiny as escape from Earth. This is a veritable secular eschatology. Famous science fiction writer Ray Bradbury stated spiritedly, "we're heading back to the stars . . . Our children's children will be immortal. That's what space travel is all about."[84] Arthur C. Clarke believes that space represents the mysterious that draws us to it, "only through space flight can mankind find a permanent outlet for its aggressive and pioneering instincts. The desire to reach the planets is only an extension of the desire to see what is over the next hill . . . Perhaps one day men will no longer be interested in the unknown, no longer tantalized by mystery. This is possible, but when man loses his curiosity, one feels he will have lost most of the other things that make him human."[85]

Historian of Technology Professor David Noble argued for corollaries between past belief in *ascension of the saints* and *modern space flight*. From the very beginning of the modern age the mystic Tomaso Campanella believed in a heavenly paradise on the moon. His contemporary, astronomer Johannes Kepler believed traveling to the moon was a way of escaping earthly turmoil. Founding father of the Royal Society Bishop John Wilkins similarly asserted that paradise exists on the moon and as soon as science perfects the art of flying colonies will be established there. Jules Verne used allusions to, "Ascension" for space flight and was the great inspiration behind the twentieth century's push to reach the moon, *the Space Race*. Flight became associated with Christ's Ascension and achieved the status of metaphor for spiritual ambitions.[86] As Historian Michael Sherry recognized, flight is a symbol of religious ascension, the realization of lofty goals. "Therefore . . . [air flight] was uniquely capable of stimulating fantasies of peacetime

---

[83] Black Sabbath, "Into the Void" *Master of Reality* (Warner Bros., 1971).

[84] Ray Bradbury, quoted in Ferkiss, *The Future of Technological Civilization*, 277.

[85] *Arthur C. Clarke, The Promise of Space (New York: Bantam 1968),* 292-93.

[86] David F. Noble, *The Religion of Technology: The Divinity of Man and the Spirit of Invention* (New York: Knopf, 1997), 120.

possibilities for lifting worldly burdens, transforming man's sense of time and space, transcending geography, knitting together nations and peoples, releasing humankind from its biological limits. Flight also resonated with the deepest impulses and symbols of religious and particularly Christian mythology—nothing less than Christ's ascension. Its realization, then served as a powerful metaphor for heavenly aspirations and even, among the literal-minded, as the palpable vehicle for achieving them."[87] Historian Walter McDougall similarly commented on the symbolic meaning of *Sputnik I* (1957) as herald of the Space Age and the coming transformation. "To some it was the newest and most spectacular evidence of mankind's irrepressible, questing nature." [88] Others believed that the promise of space technology and its consequences for Earth would forever alter man's nature by nurturing global consciousness through material abundance and perfect humankind.

Rocket scientist Wernher von Braun also believed it is, "Man's destiny" to reach the moon and other planets in order to preserve the, "life spark."[89]

Space flight operates as cultural myth for peace, a utopian new beginning for humanity, deliverance and immortality; meaning humanity's survival depends on space exploration, our destiny resides in space not earth. According to this cultural directive it is God's will for humanity to spread throughout the universe and fill it just as we did on Earth as co-creators with God. As beings made in God's image, we are expected to learn all we can of the creation.[90] Thus, "Teleology" as Lutheran theologian Carl Braaten declared, "may be seen as secularized eschatology."[91]

The Eternal Cosmos

THE COSMOS IS ALL THAT IS OR EVER WAS OR EVER WILL BE.

Space exploration as a central theme in science fiction provided the necessary imagination for technological expansion into infinity. The Copernican Revolution in the early modern period opened the heavens to dreams of exploration by means of other worldly flight. In the modern world view eternity is no longer the domain of God above the levels of heaven as in the pre-Copernican (Ptolemaic) system. Instead eternity resides in our earthly temporal space. Eternity, Earth and Space are now identical with the material universe. Carl Sagan spoke for most that hold a materialist cosmology when he stated rather glibly, *"THE COSMOS IS ALL THAT IS OR EVER WAS*

---

[87] Michael S. Sherry, *The Rise of American Air Power: The Creation of Armageddon* (New Haven, CT: Yale University Press, 1987), 2.

[88] Walter A. McDougall, *The Heavens and the Earth: A Political History of the Space Age* (New York: Basic Books, 1985), 4.

[89] Wernher von Braun quoted in Noble, *The Religion of Technology,* 126.

[90] Noble, *The Religion of Technology,* 115-142.

[91] Carl E. Braaten, *The Future of God: The Revolutionary Dynamics of Hope* (New York: Harper and Row, 1969), 29.

OR EVER WILL BE.'[92] Heaven has been brought down to Earth and now it must be conquered and filled. Levels of heaven separating Earth and eternity no longer exist. Earth abides in eternity separated from the rest of infinite space by an atmosphere that can be surmounted through rocket technology.

The traditional notion of the eternal divine is transposed into temporal humanity, into a transcendental subjectivity. The transcendent realm becomes mixed into immanent categories. The dream of space flight serves as an excellent example of the modern transposing of Heaven and Earth. In the traditional model of the universe the heavens were the abode of angels, saints and immortal spirits attainable only by ascension. Ascension is the completion of salvation in the Christian tradition. Space flight equals the new ascension to glory, mankind's reach for immorality amongst the stars. Theologian Paul Tillich described how the classical archetypal symbol of space and time, the vertical line (transcendence) has collapsed into the horizontal line (immanence) so that the vertical no longer exists today as it did in the pre-modern world. Space exploration represents the preliminary last step of the horizontal line over the vertical that began in the Renaissance.[93]

The aim of the horizontal line or *radical immanence* is the control of nature by mathematics and calculating reason, projected into outer space fulfilling human destiny. Tillich said, "The heavenly utopia as traditionally imagined was transformed into an earthly one. 'Utopia' means something hoped for which so far has no place in reality. The hope now became, not fulfillment above, but in time and space. This lifted the importance of technology, compared with pure science, far above what was possible in classical Greece and the intervening periods."[94] The modern mind is essentially a transformation from world transcendence in the pre-Copernican system to world transforming in the post-Copernican world. "One of the most important transformations is the turn from Greek contemplative and the medieval self-transcending ideals of life to the active, world-controlling and world-shaping ideal. This implied a high valuation of technical sciences and the beginning of that fertile interaction between pure and applied sciences which immensely contributed—and is still doing so—to the fast development of both of them."[95] Transcendence, the vertical line, is found within the horizontal dimension giving religious relevance to all modern scientific and technological endeavors. This is especially prominent in space exploration as a symbol for human destiny. The technological imperative is a form of

---

[92] Carl Sagan, *Cosmos* (New York: Random House, 1980), 4.

[93] Paul Tillich, "The Effects of Space Exploration on Man's Condition and Stature" in *The Future of Religions* (New York: Harper and Row, 1966), 4.

[94] Paul Tillich, *The Irrelevance and Relevance of the Christian Message* (Cleveland, OH: The Pilgrim Press, 1996), 29.

[95] Idem, "The Effects of Space Exploration on Man's Condition and Stature," 40.

religious transcendence or human ascendancy to divinity via technological progress; it has become, "the very measure of modern enlightenment . . . rooted in religious myths and ancient imaginings . . . the defining mark of modernity."[96] Conquest of infinity as prime symbol encapsulates the essence of Western Soul according to Social Philosopher Oswald Spengler, "the outward and upward-straining life-feeling—true decedent, therefore, of the Gothic—as expressed in Goethe's Faust monologue when the steam-engine was yet young. The intoxicated soul wills to fly above Space and Time. An ineffable longing tempts him to indefinable horizons. Man would free himself from the earth, rise into the infinite, leave the bonds of the body, and circle in the universe of space amongst the stars."[97] Time and space are one, theoretically; technological acceleration is racing towards a final crescendo.

---

[96] Noble, *The Religion of Technology*, 3, 9.

[97] Oswald Spengler, *The Decline of the West*, Vol. Two trans., by Charles Francis Atkins (New York: Knopf, 1938), 503.

CHAPTER TWO The 21st Century as Technological Utopia

Nineteenth century Victorians rejoiced to see our day and saw it with joy unspeakable. In the mind's eye, they projected a collective dream: the city of the future. A popular Victorian theme found throughout exhibitions, art, novels and stories. They peeked into time, looking just over the horizon, leaving the 20th century behind. They reveled in the 21st century technological utopia. They envisioned self-propelled cars traveling at mind scrambling speed. They wished for colossal building complexes with chrome spires piercing the sky, towers ascending to heaven. A magical world where ships lighter than air fly to and fro around the moon, gargantuan supermarkets overflow with goods, food and clothes. In their future people no longer trade with money but government issued credit cards: a haven for Corporate Capitalism. They saw the age of plenty; workers secure through endless industrial production, safe from grinding poverty, hunger and rampant pandemics. They dreamed of immaculate white cities enlightening dark Satanic Mills. Where machines liberate women, love is free, and humanity transcends the natural order of things. They imagined a new world order of universal peace and global prosperity. That vision ended, exploded with the atom bomb, or did it just begin?

Space Exploration as Cultural Metaphor

Science fiction does not predict the future but inspires it. Popular movies, short stories and novels reflect widespread hopes and fears and often have an uncanny ability to prefigure future technology. Yesterday's science fiction as tomorrow's science fact is over simplifying, but clearly a demonstrable cultural pattern exists. Visionaries have all set their gaze to a better tomorrow, a new world providing the inspiration for later generations to strive after the impossible. Visions for the twenty-first century utopia and space exploration tied inextricably into each other. The twenty-first century was expected to be the time of technological miracles, such as unlimited food production, disease eradication, instant communication, rapid transportation and especially space exploration. Christophe Canto and Odile Faliu explained in their marvelous compendium on the mythical history of the year 2000, "The expectation was that by the dawn of twenty-first century the solar system would already be conquered and mapped, and the galaxy as a whole would be preparing to welcome its discoverers."[98] Space exploration as cultural metaphor became associated with an infinite future for perfected humanity. After the Earth was explored and mapped the next logical step for the human spirit must be outer space. "The project was one of driving back—at least at the conceptual level—the limits of the knowable; to embark on the conquest of mathematical and temporal infinities. To fire flares into the mists of the unknown."[99] Arthur C. Clarke speculated, "The advent of space travel will produce an expansion of scientific knowledge perhaps unparalleled in history."[100] Historian Joseph Corn and Museum Curator Brian Horrigan in their fabulous museum exhibition: Yesterday's Tomorrows, educate us on how past generations of Americans envisioned their future. "According to popular culture, the future is going to take place in a single vast locale—outer space."[101] Movie critic Allan Frank added, "Space travel is the most perennially popular theme in science fiction, and its literary progenitors."[102] The goal of exploration and development must continue even after the earth has been fully mapped and charted. The space genre appeared just when the North American Frontier was about to close and opened a new and infinite horizon for human progress and conquest of the unknown.

Le Voyage Dans La Lune

The literary genre of space travel begins with Jules Verne's science fiction novels in the 1860's: *From the Earth to the Moon* and *Around the Moon* and continued later with the publication of H. G. Wells' novel *The First Men in the Moon* (1901). The genre was adapted to film from the very beginning of movie history with Georges Méliès' famous master-piece: *Le voyage dans la lune, A Trip to the Moon*, (1902). Space genre reflected the excitement and danger of potential discovery that space travel was believed to offer. Outer space enabled the modern story teller to transcend the limits of earthly time and picture a glorious destiny through the technological conquest of the cosmos. Professor of Public Affairs Howard McCurdy expounded upon the impact of past imaginary futures had on present day cold-hard reality. "In most cases, images do not duplicate reality. They may anticipate reality, but they cannot capture it exactly," he said, "Some images are so fantastic that they can never come true, although that does not necessarily lessen their influence. Notwithstanding their veracity, what matters is the capacity of

---

[98] Christophe Canto and Odile Faliu, *The History of the Future: Images of the 21st Century* trans., Francis Cowper (Paris: Flammarion, 1993), 120.

[99] Ibid, 122.

[100] *Arthur C. Clarke, The Promise of Space (New York: Bantam 1968), 293.*

[101] *Joseph J. Corn and Brian Horrigan, Yesterday's Tomorrows: Past Visions of the American Future (Baltimore: The John Hopkins University Press, 1996), 28.*

[102] Allan Frank, *The Science Fiction and Fantasy Film Handbook* (Totowa, NJ: Barnes and Noble, 1982), 185.

images to motivate behavior. Myths rooted in works of fantasy, such as fairy tales, can transmit values and shape popular beliefs in powerful ways."[103]

The reality of today's space exploration was sparked by yesterday's popular imagination that mankind can break Earth's gravity, which was thought impossible at the turn-of-the-twentieth century. Propelling literati as prophets. *Le voyage dans la lune* was the first science-fiction film ever made, yet the only scientificly accurate projection proved to be the splashdown scene. People traveling to the Moon in, 1902, although an impossible fantasy, appeared a little more plausible in the form of a gigantic cannon that would propel a capsule. Méliès borrowed this element of the movie from Verne's novels, which was the first science-fiction attempt to reach the moon with the exaggeration of the given technology at the time. Although not feasible, the idea that a powerful explosion could propel a craft into orbit served as inspiration for the first generation of rocket scientists including the Russian Konstantin Tsiolkovskii, the German Herman Oberth and the American Robert Goddard, all imagined space flight first, then set about to invent the means that made it possible.[104]

Modern prophets were unscientifically inspired to push the limits of what was percieved as impossible, inspiring the next generation to create the science and technology necessary to accomplish their vision, making the impossible possible. The impossible dream was finally realized July 20th, 1969, "they will be able to do anything they imagine, nothing will be impossible for them."[105] Film and space exploration were born alike, an inseparable Twain, a twin creation, a symbiotic complimentary of opposites, as movie fantasy inspired its own realization through the scientific endeavor.

Frau in Mond

Another influential film on the development of space technology was *Frau in Mond, Woman in the Moon*, Fritz Lang, (1929), generally considered one of the first serious science fiction films that depicted a realistic rocket count down and blast off. Although long and boring by today's standards, the movie has a spectacular rocket the *Friede* designed by early rocket scientist Herman Oberth which was a multi staged, liquid fueled rocket launched while submerged in water. The submerged lift off was incorrect, but NASA did have plans to launch rockets in this way. The plot revolves around an ostracized professor who theorized the moon was filled with vast gold deposits which becomes the motivating factor for space travel. The movie was incredibly influential on German rocket scientists at the time and in the development of the V-2 rocket by von Braun at the Peenemünde facility, which adopted the

---

[103] Howard E. McCurdy, *Space and the American Imagination* (Washington DC: Smithsonian Institution, Press, 1997), 4.

[104] Ibid, 13-17.

[105] Genesis 11: 6b.

Frau in Mond logo placed at the base of the rocket. McCurdy related the movie's importance. "The movie allowed Oberth and his associates to spread the gospel of space flight among the German intelligentsia. As one of Oberth's disciples explained, 'The first showing of a Fritz Lang film was something for which there was no equivalent anywhere . . . It is not an exaggeration to say that a sudden collapse of the theater building during a Fritz Lang premiere would have deprived Germany of much of its intellectual leadership at one blow.'"[106] Others asserted, "Many historians agree that Lang's feature had a real effect, helping to spark the imagination of real scientists to make the fantasy—moon travel—to come true."[107]

Destination Moon

Frau in Mond served as a precursor to *Destination Moon*, Irving Pichael, (1950). Although the film appeared subdued and dated with a lack of action, it showed an uncanny realism in depicting the lunar landscape, space suits and rocket flights 19 years before the first lunar landing, generally regarded as the first realistic account of space flight in film. The movie reasons that America must be the first on the moon before her enemies to maintain political and military supremacy. If the government is not going to fund the mission, then private industry must take up the slack. The movie, "was meticulously concerned with authenticity, with logic, with wonders of scientific achievement, and . . . was prophetic of the manned Apollo Moon flights."[108] The movie was a clear case where life imitated art. "Interestingly, when the U. S. astronauts made the first moon walk in 1970, Destination Moon was used by the TV networks as almost a blueprint for what was to come."[109] Another reviewer observed that the movie, "made the idea of space travel not only plausible but fascinating . . . *Destination Moon* approached the subject with vividness and believability, and permanently changed the average moviegoer's mind about space travel."[110] As a realistic treatment *Destination Moon* became influential on film director Stanley Kubrick's master-piece *2001: A Space Odyssey* (1968) and its insertion of animation as a pedagogical device was adapted by Steven Spielberg's *Jurassic Park* (1993).

Beyond the Infinite

No greater example exists of a film that associated the 21st century with space exploration than *2001: A Space Odyssey*, which was a major turning point in the history of film that propelled the science fiction genre to

---

[106] McCurdy, *Space and the American Imagination,* 15.

[107] James R. Parish and Michael R. Pitts, *The Great Science Fiction Pictures* (Metuchen, NJ: The Scarecrow Press, 1977), 142.

[108] Vivian C. Sobchack, *The Limits of Infinity: The American Science Fiction Film*: 1950-1975 (New York: A. S. Barnes and Company, 1980), 21.

[109] Parish and Pitts, *The Great Science Fiction Pictures*, 95.

[110] Bill Warren, *Keep Watching the Skies!* (Jefferson: NC: McFarland, 1982), 2.

respectability. "*2001* [was] the movie that changed science fiction from low-budget exploitation fare to a subject worthy of intelligence and an artistic sensibility."[111] Film analyst Keith Booker informed us that *2001*, "resurrected science fiction film at a time when the genre had largely faded from view . . . Brilliant, thought-provoking, and visually stunning, *2001* brought new credibility to the science fiction film as an art form, demonstrating the ability of film as a medium to produce precisely the kind of mind-expanding images that readers of science fiction literature had long imagined but had never before seen on the screen."[112] *2001* represented science fiction movie come of age. "*2001* was a clear sign of things to come: hereafter, movie makers everywhere would have to take the SF [science fiction] genre seriously. More than any other picture, *2001* gave science fiction film respectability, and a sense of growing maturity."[113] Central to the movie was a rogue computer, reminiscent of Frankenstein's monster, the HAL-9000—*Heuristically* programmed *ALgorithmic* computer—representing the highest achievement in advanced technology. *2001* broke new ground in cinema with its use of sound and visual impact. The movie won the Oscar for special effects and was nominated for best director, original screenplay and best set direction. *2001* was ranked twenty-second on the 1998 American Film Institute's list for all-time great films, and the second highest ranked science fiction film in history, trailing only behind *Star Wars*, which ranked fifteenth of all time. Booker remarked that the special effects of the movie made, "before computerized imagery revolutionized the entire field, continue to impress; the way in which Kubrick's camera seems almost lovingly to caress the technological artifices it films represents a powerful pro-technological statement,"[114] despite HAL's malfunctioning. This is a misinterpretation of the film. As Booker expressed, "When there are themes, they seem more metaphysical than political."[115] The fact that HAL appears, "the most 'human' of all the characters in the film"[116] only underscores dehumanization in technosecular society. Michel Ciment reported that Kubrick was fascinated by the cybernetic belief in human replacement by machines. "The aim of cybernetics is to replace man with machinery for all routine work, for all operations that are mechanical and intermediary."[117]

---

[111] Ibid.

[112] M. Keith Booker, *Alternate Americas: Science Fiction Film and American Culture* (Westport, CT: Prager, 2006), 75.

[113] Douglas Menville and R. Reginald, *Things to Come: An Illustrated History of the Science Fiction Film* (New York: New York Times Books, 1977), 146.

[114] Booker, *Alternate Americas*, 82.

[115] Ibid, 83.

[116] Ibid, 86.

[117] Michel Ciment, "The Odyssey of Stanley Kubrick: Part 3: Toward the Infinite—2001" in William Johnson, Ed. *Focus on The Science Fiction Film* (Englewood Cliffs, NJ: Prentice-Hall, 1972), 134.

The character with the most personality and the one we empathize with is HAL the supercomputer. A direct connection between Cartesian thinking that man is a machine and *2001*'s reductionism serves as a critical theme that emerges when Kubrick reduces people to machines and humanizes machines. This fact emphasized Kubrick's point that homo sapiens approach exhaustion as a species and that the new machine intelligence represents a genuine threat as evolutionary replacement. The film was broken into four Acts. Act I, *The Dawn of Man* shows human evolution through late primate contact with an extraterrestrial black monolith that enables the hominid to use technology in fashioning a weapon from a femur bone—technology is murder. In Act II, *untitled*, the movie cuts to the 21st century and shows the marvels of space flight and a second encounter with the same monolith on the moon which emits a radio signal to Jupiter. Act III, *Jupiter Mission* presents the spaceship *Discovery* en route to Jupiter and the last Act shows one man's transfiguration after his encounter with the same monolith in space, aptly titled *Jupiter and Beyond the Infinite.*

The movie portrayed the culmination of the space theme with an ancient twist reminiscent of the pre-Copernican world view. In space mankind will reach its destiny through technological progress but will eventually transcend the need for technology as it culminates in a higher spiritual ascendancy. Technology is necessary for humanity to evolve from its primitive condition and we can surmise that it is a divine gift since the great evolutionary jump came after early man's contact with an extraterrestrial black monolith. However, the next stage of evolutionary development also follows contact with the same monolith in space resulting not in technological advancement, but human divinization.

Technology in *2001* reflected humanity's murderous nature. The hominid achieves enlightenment through a short period of trial and error by poking around in a bone field. The bone was quickly weaponized as one hominid, bludgeons another in a struggle for survival. The HAL-9000 seen only in Act III, absent the extraterrestrial monolith serves as the model for all killer computer movies that follow. "The notion of runaway computer that turns on its programmers is by now a virtual cliché . . . though it was not quite as familiar in 1968 as it is now."[118] HAL controls all operations of the ship and considers himself *perfect* and *inerrant.* He reproduces the activity of the human brain with greater speed and accuracy. HAL is essentially glorified as a perfect sentient entity and the obvious replacement for humanity since he can operate the ship autonomously. Only he knows the mission to encounter extra-terrestrial life. After HAL discovers the crew's plan to disconnect him because of a confirmed error in calculation, HAL murders three hibernating astronauts by disconnecting their life support, attacks a fourth with a shuttle

[118] Booker, *Alternate Americas,* 81.

on a spacewalk and strands a fifth by refusing him reentry to the ship. HAL feels he can accomplish the mission without humans who are prone to error. Their elimination is the most logical and surest way to ensure success.

The fact that HAL believes himself a suitable replacement for people coincides with the main point of the movie, which is that humanity will soon transcend its current dependency on technology, which has always been murderous and violent since the first primate weaponized a bone to our highest technological achievement in space travel and artificial intelligence. Daniel Dinello commented on the famous bone to spaceship scene connecting the first and second act, a match-cut spanning, four million years. "The victorious man-ape then heaves the bone into the air. Kubrick then cuts to a spaceship—the most powerfully symbolic graphic match in the history of cinema."[119] Commander Bowman narrowly escapes death in his struggle against a competing intellect after disconnecting HAL, only to die of old age and reborn as star child after his encounter with the monolith in Jupiter space. The Messianic star child represents humanity's next evolutionary step at the brink of extinction before a greater evolutionary intelligence represented by HAL. Kubrick confirmed this meaning when he said concerning Bowman, "He is reborn, an enhanced being, a star child, an angel, a superman, if you like, and returns to earth prepared for the next leap forward in man's evolutionary destiny."[120] The star child is the first of a new species representing a spiritual rather than technological evolution. "The human succeeds by employing determination, creativity, and mobility—things the computer lacks. Kubrick's vision reveals technology as a competitive force that must be defeated in order for humans to evolve."[121] Kubrick's master-piece revives old ideas of human glorification but can also be read in a modern sense of humanity finding its destiny in space through technological development.

The City of the Future

Let us create Mega City One with a huge tower
ascending into heaven and our name will be GREAT.
Genesis 11: 4

Visions of the future are dominated by technological reconstruction of habitation and landscape in the form of megalopolis. The city of man rises over the horizon devouring everything in its path. The American quest for New Jerusalem was usurped and recast by a transformation that turned the City of God on its head, "with the rise of the great city and the machine, the spiritual basis for ideal communities [millennial venture] was gradually

---

[119] Daniel Dinello, *Technophobia! Science Fiction Visions of Posthuman Technology* (Austin: University of Texas Press, 2005), 98.

[120] Stanley Kubrick, quoted in Thomas A. Nelson, *Kubrick: Inside a Film Artist's Maze* (Bloomington, IN: Indiana University Press, 2000), 133.

[121] Dinello, *Technophobia,* 99.

supplanted by more mundane values. A new image surfaced—the secular city of the capitalist future."[122] Cosmopolis or world-city is not utopia but presents the spiritual and organizational capital of a civilization. Classical Athens, Renaissance Florence, Ancient Babylon and Rome are instances of historical world-cities. Contemporary examples would be Paris, London and finally New York City, the crossroads of the world, but to be more precise Megalopolis is on the Eastern Seaboard of the United States, the concentration of urban sprawl from Boston to Washington DC occupies the contemporary world-city honors. A world-city represents, "the quintessence of a civilization, the gathering of all its vital human resources into a living organic unity. A Cosmopolis is not utopia; it is not the best of all possible worlds, but the boundless community of the best in the world . . . Cosmopolis is simply the world in a state of optimal integration."[123]

New York City from the 1880's to turn-of-the-century, dominated visions of the City of the Future, densely populated with towering skyscrapers and aerial bridges, "the world Manhattanized" appeared to Victorians as, "the very essence of modernity and the crucible of the world's urban future."[124] Individuals had become an amorphous biological mass appendage to the machine, squeezed into cubicles, tenements and Housing Projects in futuristic visions of the secular city. Professor Clarke insightfully said, "The state of future society was already apparent in the [urban] conditions of 1890."[125] The City of Tomorrow was largely the present, turn-of-the-century, urban landscape extrapolated to the final degree, full of wonderful new flying and driving machines, towering Babel's, but little to no people. Chicago's Colombian Exposition (1893) served as virtual blue print for 20th century urban development. The "White City" illuminated by electric candle captured Victorian imagination. Likewise, at Futurama in New York's World's Fair (1939) viewers watched in awe as the world of 1960 sped by on miniature freeways and scale modeled skyscrapers.

By the 1920's the paradigm for the great city shifted west to Los Angeles. Architect Richard Neutra's drawing *Rush City Reformed* (1928) was hailed as "the first futuristic vision to recognize the prophecy latent in Los Angeles."[126] The name Rush City foresaw the speedy tempo that American life would become in the second half of the 20th century. Rush City was designed for the efficient movement of traffic, buildings sit adjacent to rapid transit systems, and the entire city accessible by automobile, yet, "eerily devoid of human

---

[122] *Corn and Horrigan, Yesterday's Tomorrows,* 35.

[123] W. Warren Wager, *The City of Man: Prophecies of a World Civilization in Twentieth-Century Thought* (Cambridge, MA: The Riverside Press, 1963), 15.

[124] *Corn and Horrigan, Yesterday's Tomorrows,* 35.

[125] Clarke, *The Pattern of Expectation*, 189.

[126] *Corn and Horrigan, Yesterday's Tomorrows,* 39. 54.

beings."[127] The Modern Movement in architectural design along with Bauhaus and International Functionalism—form follows function—rose to prominence in the second half of the 20th century and left a legacy of architectural iconoclasm (destruction of all traditional societal images) which rejected the past and embraced the skyscraper as predominant in their vision of the Metropolis of Tomorrow. The United Nations building in New York City was its most famous edifice. "Nothing of the present city would remain; the brave new world of the future would require a tabula rasa. The modern vision was futurism as iconoclasm."[128] Prominent among the Modernists was Swiss architect Le Corbusier, whose vision of the city of the future as a machine for living represented in his drawing, *A Contemporary City for Three Million* (1923) dominated urban planning in the latter 20th century. Le Corbusier depicted the city of the future with huge towers and open court yards below, which presaged Chicago's infamous Cabrini-Green Projects and St. Louis' Pruitt-Igoe housing development both of which have been declared uninhabitable and subsequently demolished. As prophetic as their vision was the Modernists remained, "frighteningly megalomaniacal and dehumanizing. Technology ran far ahead of human necessity and understanding, overwhelming everything, apparently replacing all mundane facts of human life."[129] Architects became the self-appointed social engineers of the future and High Priests of the Machine through designing correct physical environments. They believed human behavior would follow. The square box absent all personality and character or ornament, such as traditional culture, regional background or landscape might afford was their basic design, which they believed to be the application of the scientific method to art. As Le Corbusier glibly pronounced there would be, "one single building [design] for all nations and climates."[130] This colossal uniformity in buildings and their integration with transportation technology was central to the Modernist vision of the great city, yet conspicuous by its absence was the human presence. The city of the future as machine for living morphed into the living dead machine.

Futurist visions of the city take finally shape in the 1960's and 70's with images of the megastructure, "a single vast, unified structure, encompassing all areas of human activity."[131] The city as megastructure replaced the natural environment entirely as a colossal planetoid machine floating above the landscape more reminiscent of a huge spacecraft than a city, with the clarion call to leave Earth and conquer unknown space.

---

[127] Ibid.

[128] *Corn and Horrigan, Yesterday's Tomorrows*, 39.

[129] Ibid.

[130] Le Corbusier quoted in Louis Hellman, *Architecture for Beginners* (New York: Writers and Readers Publishing, 1988), 151.

[131] *Corn and Horrigan, Yesterday's Tomorrows*, 54.

Intermezzo

WORDS are symbols representing reality outside themselves but not exclusive of that reality. Only by the word do we have the power to bring imagination to life.

MACHINHEAD

is one such word: half man, half machine, fully God. Well, saying it doesn't make it so. If we believe in what we say, it does. What the heart believes the mouth confesses. Just as we must see before we desire, so we must listen before we believe. Faith comes from hearing the word of God, which is so simple that its greatest practitioners are children. The word of God is the sublime power of the Almighty who only need speak to create. He who said, "Let there be light!" Also said, "I make all things new!" We walk by faith not by sight. Faith is confident hope that transforms invisible truth, "evidence of thing not seen" into visible, tangible reality. Faith makes the impossible possible. We have the power to create gods in our own image and likeness. We give existence to Machinehead or not.

Part II

Knowledge

The future of the present is in the past.

CHAPTER THREE Machinehead Rise to Glory

Space exploration offers only one potential future. Space mythology exists to bolster acceleration on Earth. Given our choice of futures why should we settle for the rarefied environment aboard the space station. Once we remove the determinist premise from our thinking, we can turn our attention back to earthly concerns like ending war and curing poverty. Our destiny is not in the stars, but in each other. Once we realize this, only then will we inherit the Earth. Instead we remain enthralled with the machine and advantages in war. A technological imperative has taken over our lives. We persist in thinking that there is yet a future waiting us in the vast reaches of outer space. Dehumanization starts with removing the social animal from humanizing influences of wilderness, family and tradition. The eternal cosmos replaces the eternal God, and we are subsumed into nature as object of science rather than subject. The 21st century remains a metaphor of hope from past centuries. We live the reality of other's dreams.

The Cosmic Machine

What did you dream?
It's alright we told you what to dream
. . . so welcome to the machine.[132]

The universe is a great machine, swirling galaxies within galaxies according to modern cosmology; a notion initially sketched by Rene Descartes (the father of modern philosophy) then fleshed out in the genius of Sir Isaac Newton, who bequeath his name to the intellectual superstructure—the *Newtonian World Machine*—that gives rise to that planetary entity, I am here calling, "Machinehead." Historian of Ideas John Randell elaborated, "The world is a vast perpetual motion machine, and every event in it can be deduced mathematically from the fundamental principles of its mechanical action."[133] Here we find the origins of what evolutionary theorists call, "Anthropocene"—the human epoch—mankind's journey from Classical

---

[132] Pink Floyd, "Welcome to the Machine" *Wish You Were Here* (Harvest, 1975).

[133] John Herman Randell, Jr., *The Making of the Modern Mind: A Survey of the Intellectual Background of the Present Age* (New York: Columbia University Press, 1976), 253-281.

pawn in the universe to Medieval knight of faith traveling the royal road of art and science to Modern human divinity.

"Anthropocene," means technosecular civilization is a force equal to nature with power to determine the fate of the earth. "Anthropocene," then, is a misnomer, since analysts, experts and the consuming public, unanimously agree, except this writer that technological acceleration cannot be stopped. It is a social dynamic out of control, functioning *like* a parasite on its anesthetized host, i. e. collective human agency. We, then, cannot rightly call this process, "the human epoch." However, *Technopocene*—the technological epoch—in which technology is its chief Characteristic appears justified as appropriate designation for our collective surrender to the ever-advancing strides of Machinehead. This is evident from that fertile joining between modern cosmology and industry, whose illegitimate love-child we know as super exponential population growth, relentlessly ravaging Mother Earth.

It is mankind's purpose to reintegrate himself into the machine as one flawless operation. Professor Floyd Matson tells us that the consolidation of organic nature with the inorganic material has always been the teleological purpose of Modernism since the seventeenth century. "The new science promised the transformation of man himself, along with all life, into the measurable and manipulable working parts of the great machine. If this had not been precisely the dream of Descartes, it was to become the all– absorbing vision of the post-Cartesians, brought to its fullest expression by Pierre Simon de Laplace," Matson expounded. The twenty-first century is heir to the early modern vision of a cosmic intelligence of earthly origins that will transcend us. A mind unbounded from language restraint, bodily limits, and local customs and beliefs will be able to incorporate all knowledge and power into one indissoluble unity, to act as judge over the affairs of mortals. Matson elaborated further that this living machine would possess, "the pansophic vision of perfect prediction [All Seeing Eye], exact measurement and absolute certainty. A superhuman intelligence . . . one which could know at a given moment all the forces by which nature is animated and the respective position of the entities which compose it, 'would embrace in the same formula the movements of the largest bodies in the universe and those of the lightest atom: nothing would be uncertain for it, and the future, like the past, would be present to its eyes.'"[134] This All Seeing Eye encircling the globe threatens us with a synthetic replacement.

---

[134] Floyd W. Matson, *The Broken Image: Man, Science and Society* (Garden City, NY: Anchor Books, 1964), 11; Pierre Simon de Laplace, *Traité de Probabilité* (1886). http://www.universalis.fr/encyclopedie/pierre-simon-de-laplace/5-la-theorie-des-probabilites/

Cyborg(s) In Space

Mars Needs Women

Wherever the survival of humanity is threatened we find the work of Satan. Jesus said, "The thief comes only to steal, slaughter and decimate."[135] In the previous century that was Fascism, then Super Fascism known as Mutually Assured Destruction during the Cold War. Today, Satan hides behind the ascendancy of the technological world order; mankind's incorporation into the cosmic machine creating a single planetary being that will wield total systems control over the lives and destiny of everyone and everything on earth: *The Cyborg*. Jesus equated the forces of Satan with those of Man worship when he said to Peter, "Get behind me Satan! You are an obstacle to me. You are not thinking as God does, but as human beings do."[136] Berdyaev minced no words when he emphatically proclaimed, "A totalitarian state [condition] is the kingdom of Satan."[137] Theologian Karl Rahner proposed that, "The Antichrist is the embodiment of all those historical forces hostile to God which are under the control of man."[138]

Machinehead is what I call the technological idol, the planetary creature taking shape in the convergence of human and computer intelligence, one language—mind—one world, the *global cyborg*. *Cyborg* is a hybrid word short for *cybernetic organism*, coined in 1960 by two research scientists, Manfred Clines and Nathan Kline in the emerging field of cybernetics, published as a journal essay, "Cyborgs and Space" in *Astronautics* based on a lecture, "Drugs, Space and Cybernetics" given to the Air Force School of Aviation Medicine. *Cybernetics* is the field of machine psychology that studies how machines think and communicate. *Organism* is an, animate being. A *cybernetic organism* or *cyborg* is the combination of man-made machine and God-made creature. A cyborg is a living machine, whose biological components are adapted to its non-biological machine parts. The cyborg assimilates biology and chemistry with inorganic physics and technology joining natural processes with artificial systems in an ultimate *all-in-one* transfiguration or apotheosis. The Roman Imperial Cult serves as the best example of past human ascendency to divinity. The deification of leaders in Rome was the work of the masses, not government itself, "the apotheosis of the Caesars, as well as of all tyrants and dictators since, was the work of the people, the poor, rather than the senate, which was always sceptic and little

---

[135] John 10: 10.

[136] Matthew 16: 23 (NAB).

[137] Nikolai Berdyaev, *Slavery and Freedom* trans., by R. M. French (New York: Charles Scribner's Sons, 1944), 149.

[138] Karl Rahner, *Dictionary of Theology* New Revised Ed. (New York: Crossroads, 1981), 19.

inclined towards mystic beliefs."[139] Singularitarians expect the arrival of *the Great Symbiote* by mid-century, *circa* AD 2045.[140]

We will change ourselves through the process of emulation at a global scale by terraforming Earth to resemble the conditions of a space station. Futurist architect and apologist for *the geodesic dome* R. Buckminster Fuller declared, "Reform the environment—not man,"[141] by which he means that through restructuring the physical environment human nature would inevitably follow. Change the environment to change man. Earth's metamorphose is the goal of Cyborgism, which wants to turn our fertile green and ocean blue planet into a metal megastructure that will look something like life aboard the Death-Star. The Cyborg was originally conceived to transport Earth's fragile ecosystem through the rigors of outer space; now, rather than change the Moon into Earth, Earth will turn into the Moon: a space station for the liberation of Transkind. This planetary megastructure will simulate Earth's natural environment, brazenly, purporting its superiority to the original: "Spaceship Earth."[142] Literally!

Astronauts sealed inside their air-tight capsule afford the best analogy to our future imprisonment; reintegration with nature through the machine will remake human physiology and Earth's ecology in the image of artificial infrastructure. All life on Earth will utterly depend on the great "technosphere," personified as the Digital Star-God, who will bestow Earth's scarce resources to all; it will resemble an electric Pharaoh ready to transcend with its *transpeople* to the next level. The cyborg is a creature built for space travel, not terrestrial habituation.[143] It will redesign humanity for space exploration and leave the God-design behind. This Posthuman race will not inherit the Earth as the meek in character are promised.[144] They will abandon it, *even destroy it* under the presumption that a new and better world waits in the deep recesses of outer space. Earth is a steppingstone to the stars.

---

[139] Nikolai Berdyaev, *The Realm of Spirit and the Realm of Caesar* (New York: Harper, 1953), 76.

[140] Lev Grossman, "2045: The Year Man Becomes Immortal" in *Time,* (VOL. 177, NO.7/ 2011), 42-49; D. T. Max, "Beyond Human" in *National Geographic* (April 2017, VOL. 321, NO. 4), 40-63; *Apotheosis*: "elevation to divine status: deification - *the apotheosis of Roman emperors*"; https://www.merriamwebster.com/dictionary/apotheosis.

[141] R. Buckminster Fuller, *Utopia or Oblivion: The Prospects for Humanity* (New York: Bantam, 1969), 320.

[142] Idem, *Operating Manuel For Spaceship Earth* (Zurich, CH: Lars Müller, 2015 [1969]).

[143] Andy Clark, *Natural-Born Cyborgs: Minds, Technologies and the Future of Human Intelligence* (Oxford University Press, 2003), 13; D. S. Halacy, Jr., *Cyborg: Evolution of the Superman* (New York: Harper & Row, 1965).

[144] Matthew 5: 5.

Lewis Mumford pinpointed the exact problem with the, "Megamachine." He declared that the goal of technological civilization is the attenuation of human nature and exhaustion of global resources through space exploration. "At the bottom of this whole effort lies a purpose that animates the entire megamachine, indeed, figures as its only viable consummation: to reduce the human organism itself, its habitat and its mode of existence, and its life-purpose [*telos*] to just those minimal dimensions that will bring it under total external control."[145] The technological order by means of entertainment denigrates life on Earth as inferior to what space travel will yet accomplish. "Space travel would be impossible without the total mobilization of the megamachine commanding to the point of exhaustion all the resources of the state," Mumford explained. Space travel is a symbol of progress and the means to total control. "Its ultimate goal . . . is to reduce this great round globe to the dimensions, so to say, of a billiard ball . . . To justify space travel, its exponents must vilify earthly life. And this is precisely what the technocratic intelligentsia do not hesitate to do, in order to justify their unconditional commitment to the megamachine."[146] This *Eschatology of Technology* removes our destiny off the earth, placing it in uncharted regions of the final frontier. Societal ambitions decide its heroes. Our culture glorifies the Astronaut as the epitome of techno-kind, supported by science fiction romance that shapes public perception with the hero conquering hostile outer limits. "'It may well be,' says Arthur Clarke. 'that only in space, confronted with environments fiercer and more complex than any to be found on this planet, will intelligence be able to reach its fullest stature . . . The dullards may remain on placid Earth and real genius will flourish only in space-the realm of the machine, not of flesh and blood' . . . The actual genius that will 'flourish only in space, in the realm of the machine,' is the genius of entropy and anti-life," Mumford added.[147]

Space exploration represents the driving cultural motif or myth in technosecular society from the Victorian era to the present. Modern History Professor Gerard J. De Groot discussed the spontaneous connection in the popular mind between all things technological, outer space and progress. "There was no grand marketing plan, but rather an ad hoc effort by hundreds of producers to latch on to the space race as a way to demonstrate how modern they were. Space, in which new things seemed to be happening every day, became synonymous with progress."[148] This probing insight depicts NASA as the premiere embodiment of technological progress, who set the marketing

---

145 Lewis Mumford, *The Pentagon of Power: The Myth of the Machine*, Volume Two (New York: Harvest, 1970), 306-311.

146 Ibid.

147 Ibid.

148 Gerard J. De Groot, *Dark Side of the Moon: The Magnificent Madness of the American Lunar Quest* (New York: New York University Press, 2006), 186.

tone during the Space Race by creating space travel as advertising theme demonstrating the power of the Corporate State or Military-Industrial Complex to craft the agenda and create the pace for global technological change in both the East and the West.[149] The nineteenth century idea of inevitable progress must eternally pursue new frontiers to conquer and explore or suffer from perfection, when science reaches its end or full potential, in which all that is to be known, discovered and invented has already finished, leading to global technological stasis or equilibrium, followed by inevitable social decline (degeneration and regression). Space exploration keeps the window of acceleration open and necessary forever.

Mission to Mars is NASA'S latest gambit and publicity campaign to free itself from the deep freeze of public disinterest in space exploration and its own petrified reputation for super redundancy. A New Space Race with the Chinese is just the political oil needed to reignite public excitement for space travel as the recipe for technological stagnation and decay sure to follow if NASA trails the East in the race to Mars, so the argument will go. Mars or Bust! Is a false choice, but it will be made to appear as though human survival depends on Mission to Mars, given the exhaustion of Earth's resources and the threatening disaster of global population growth reaching nine billion souls by mid-century. Techno-critic Jacques Ellul reminded us that the first space race to the moon failed in achieving its goals to colonize and mine its hypothetical vast mineral reserves. "The moon has been abandoned," he said, "We should not forget that the main aim of the Apollo program . . . was not simply to go to the moon but to settle there and build factories. This aim has been abandoned by the Soviets as well as the Americans. In another century or so it might be possible to exploit other planets, but the expense is enormous, and the usefulness of the fabulous exploits is nil."[150] Mumford continued his criticism of the space program in a familiar theological framework as the serpent's temptation to forbidden knowledge and power. "With space exploration the traditional enemy of God and man has already

---

[149] "The phrase *Military-Industrial Complex was first used in an American report at the turn of the 20th century* and later immortalized by outgoing United States President Dwight D. Eisenhower during his January 17, 1961 farewell address to the nation. In his speech Eisenhower cited the '*Military-Industrial Complex' as a grave warning to the American people based on his experiences of an unlimited wartime economy coupled with a political environment as witnessed during, and after, World War 2 (1939-1945) – the warning being to not let the military-industrial establishment dictate America's actions at home or abroad for such unchecked power would begin to usurp the inherent freedom found in the very fabric of our nation. The original usage appeared as Military-Industrial Congressional Complex but this was later – rather ironically – revised to exclude its reference to the American Congress."* – http://www.militaryindustrialcomplex.com/what-is-the-military-industrial-complex.asp

[150] Jacques Ellul, *The Technological Bluff* trans., by Geoffrey W. Bromiley (Grand Rapids: Eerdmans, 1990), 270.

re-appeared, in Post-Faustian form. And as of old, if one is willing to sell one's soul to him, he offers his ancient bribe—unlimited power of control, control absolute, not only over all other kingdoms and principalities, but over life itself."[151] Mumford was talking about Posthuman restructuring life on Earth, which means every living thing alive today will be irrevocably changed forever according to the dictates of market forces, genetic engineering, artificial intelligence, nano technology and the caprice of the super-rich and government agencies for the expressed purpose of transforming Earth into a giant space station—megastructure—to serve as the launching pad into the cosmic void for a Posthuman race. "To sin against the earth is now the most dreadful thing," the Prophet Zarathustra declared, and, "THE LORD GOD ALMIGHTY . . . [will] "DESTROY THOSE WHO DESTROY THE EARTH" the Apostle John revealed. Nietzsche and the Bible agree humanity and Earth are inextricably one. To abandon earth is apostasy.[152]

Machinehead: A Definition

Machinehead is a compound name that joins the words, "machine" and, "head." A machine is *a* closed *system within a system of human origins, wheels within wheels* or contrivance, a miniature cosmos made for technical efficiency in relationship to human society; it is a principle of absolute organization expanding to include the entire world. The machine is a paradigm of perfect order defined abstractly it consists in a cycle of interconnected agencies made to accomplish a particular end. The machine has significantly affected social structure. Organizational function relies on predictability that internally creates more organization. "Order demands order." [153] There is no such thing as an isolated system either in the realm of physics or social interactions. All systems relate to adjoining systems in contiguous motion dispersing chaos and formlessness. By narrowing the parameters of nature and society it refocuses life on its own technological ability to surmount one difficulty after another by dissolving natural and traditional boundaries. The machine in relation to human life is analogous to the biological phenomena in which normal cells are transformed by the spread of cancerous ones throughout the body.

The word "head," of course, refers to the *divine essence* as theology uses it to describe to the fullness of God's Trinitarian nature as in, "Godhead" or, "Godhood." This meaning suggests that the substance of God is differentiated, reflected in self-consciousness intimating plurality within singularity. Theologian W. Shedd elucidated, "Mere singleness, a mere subject without an

---

[151] Mumford, *The Pentagon of Power,* 311.

[152] Revelation 11: 18; Fredrich Nietzsche, *Thus Spoke Zarathustra: A Book for None and All* trans., by Walter Kaufmann (New York: Penguin, 1966 [1891]), 13.

[153] Roderick Seidenberg, *Posthistoric Man: An Inquiry* (Chapel Hill, NC: The University of North Carolina Press, 1950), 27, 28.

object, is incompatible with self-consciousness. And mere duality would yield only consciousness, not self-consciousness. Consciousness is dual; self-consciousness is trinal."[154] The Apostle referenced divinity known to all people through nature and the incarnation of the Word of God in Jesus Christ.[155] This distanced the work of God in the creation and human embodiment from the technological works of our hands. "Since we are God's offspring," he preached, "we should not think that the divine being is like gold or silver or stone—an image [statue] made by human design and skill."[156] Thus, technology does not reflect transcendent deity, but strictly expresses human deification of itself as a machine. *Machinehead!*

Machinehead, then, as I use the word means *the planetary Cyborg's ascension, rather, usurpation to Godhood.*[157] People best understand large conglomerates when personalized, such as, referring to the Federal government as, "Uncle Sam" so I have chosen to name this, so called, New World Order: *Machinehead.* Personification of mystery strives to master the unknown by understanding. We project our highest values, ideals and vices into divinity. Berdyaev echoed German religious philosopher Ludwig Feuerbach when he intoned, "Man creates God in his own image and likeness and puts into God not only the best in his own image but the worst also."[158] Mankind's creation of God reflects human limitations projecting the traditional social order of master-slave into the metaphysical order turning the Living God into an all-powerful master and man as his slave; concomitantly, *oriental despotism* (social norm of history) always pairs with *cosmic determinism* (metaphysical norm of history), which means God is identified with the impersonal cause and effect of the cosmos such as natural law, subjecting human freedom to servile determinism. Machinehead transposes that traditional order into a social bureaucracy giving the machine the appearance of God on Earth.

Hegel defined *Oriental Despotism* as the knowledge that only one man is free: The Emperor, Khan or Mogul. "Orientals do not yet know that Spirit—Man as such—is free. And because they do not know it, they are not free. They only know that one is free; but for this reason, such freedom is mere caprice . . . This *one* is therefore only a despot [a ruler with absolute

---

[154] William G. T. Shedd, *Dogmatic Theology* 3rd Edition (Phillipsburg, NJ: Presbyterian and Reformed Pub., 2003 [1888]), 172.

[155] "Divinity" means "Godhead" in the KJV (Romans 1: 20; Colossians 2: 9).

[156] Acts 17: 29 (NIV).

[157] Lawrence J. Terlizzese, *Into the Void: the Coming Transhuman Transformation* (Cambridge OH: Christian Publishing House, 2016).

[158] Berdyaev, *Slavery and Freedom*, 82; Ludwig Feuerbach, *The Essence of Christianity* trans., by Marian Evans (New York: Calvin Blanchard, 1855 [1841]).

power], not a free man."[159] According to Hegel history moves towards, "One Individuality" or *Singularity* in today's terms, culminating in the reign of the *despot* embodied, "essence as God . . . inseparably united with the Spirit of the State."[160]

Cosmic Determinism means man is slave to cosmic causes, such as *the natural order of things, Karma, Necessity, Fate, the law of cause and effect.* Ancient Historian F. Cornford clarified that the macro concept of *Moira* (necessity, fate or destiny) underlies all religious and philosophical belief, conjecture, and understanding in the ancient world. *Moira,* Greek literally, "part" or "allotted portion"; "from that primary meaning it is agreed that the meaning 'destiny' is derived." *Moira* is the ancient version of cosmic determinism meaning the impersonal force of the universe by which all things are ordered, serving as the rational basis for religion, morality, law and justice. Moira is a keystone idea around which the ancient world centered itself from it we derive the nature of a thing, reality and the world-soul. "It is a material continuum," Cornford expounded, "which is also alive and divine, soul and God—a substance . . . invested with mystical properties."[161] When we vehemently maintain that our technological practice is necessitated by commonsense realism, or we admit that we cannot live without our favorite devices, we give place to modern Moira. Machinehead represents cosmic determinism and oriental despotism in digital form. The eternal order of things integrates man and machine as axis point.

Diaboli Ex Machina

We have Done the Devil's Work.
J. R. Oppenheimer
(Father of the Atom Bomb)

Cosmocentricism (cosmos centered world view) has reasserted itself in the contemporary technosecular world order subjecting humanity to a renewed slavery patterned according to political despotism and cosmic determinism. Despotism means man is slave to the religiopolitical order, which believes itself to mirror the cosmic order; for example, the Divine Rights of Kings in Western tradition and Oriental Despotism in the East; the norm of despotic religious order is a reflection of the natural order, as a traditional Confucian saying puts it, "there are not two suns in the sky, there cannot be

---

[159] Georg W. F. Hegel, *Reason in History: A General Introduction to the Philosophy of History* trans., by Robert S. Hartman (Indianapolis: Bobbs-Merrill, 1953), 23.

[160] Idem, *The Philosophy of History* trans., by J. Sibree (Buffalo, NY: Prometheus, 1991), 53.

[161] F. M. Cornford, *From Religion to Philosophy: A Study in the Origins of Western Speculation* (New York: Harper & Row, 1957), 16, viii, ix.

two emperors on earth."[162] These ideas of God as impersonal force and cosmic despot Berdyaev called the, "last refuge of human idolatry."[163] The social order reads the human condition through nature and society back into the cosmic or metaphysical order leaving the world with an anti-human view of God, which is understood as absolute and impassible, reminiscent of the classical cosmocentric philosophy that Christianity replaced with an anthropocentric one (human centered world view). *Christian Anthropocentricism* was a new world outlook favorable to salvation and survival of mankind as Lord of Earth, which entrusted to us the prosperity of nature. Medievalist E. Burtt explained further, "For the Middle Ages man was in every sense the center of the universe. The whole world of nature was believed to be teleologically subordinate to him and his eternal destiny."[164] Humanity was the most important and controlling fact. Berdyaev tells us that in the modern age there has been, "a return to pagan cosmocentricity . . . the spirits and demons of nature have again arisen out of the hidden depths of the life of nature and taken possession of man. The demonolatry from which Christianity had liberated man has periodically seized upon him again."[165] Ancient pagan cosmocentricity has returned and usurped Christian anthropocentricity.

The return to cosmocentricity leaves mankind alienated from the center of all things. Earth has been cut adrift as a spec of space debris in a vast cosmic ocean, "a pale blue dot . . . a mote of dust suspended in a sunbeam," according to Cosmologist Carl Sagan. Mankind is crushed beneath the relentlessly grinding mechanism of nature. Once, Lord of Earth *imago dei* is no longer the center of cosmic order, but its slave. "The falling away of man from God has brought with it the falling away of the cosmos from man."[166] Professor Floyd Matson explained the contemporary human condition as one in which humanity finds itself exiled from the cosmic center. The great Newtonian World Machine turned against its creator, "from the center of stage—from an active part in the drama—to a seat in the audience and the passive role of spectator . . . man has disappeared from the world as *subject* in order to reappear as *object*."[167]

Technopocene (the epoch of technology) is Christian anthropocentricism that has lost its ground of being. Once rooted in transcendent principles the cosmic machine has lost touched with its subservient role in the universe.

---

[162] Quoted in Samuel P. Huntington, *The Clash of Civilization and the Remaking of World Order* (New York: Simon & Schuster, 1996), 234.

[163] Berdyaev, *Slavery and Freedom*, 83.

[164] E. A. Burtt, *The Metaphysical Foundations of Modern Science* (Garden City, NY: Doubleday, 1954), 18; Genesis 1: 26; 9: 2.

[165] Berdyaev, *Slavery and Freedom,* 101.

[166] Ibid.

[167] Floyd W. Matson, *The Broken Image*, 13.

From expressing the order of God in nature—the cosmic machine became God. Professor Randell says modern science was, "trying to reach an absolute system of truth quite independent of any limitations of the mental powers of the essentially imperfect and biological creature that man seems to be . . . they were trying to arrive at a complete and perfect understanding of the Universe that only a God possess, by methods possible for a being who is not a God but a rational animal."[168]

Experimental science, *Scientia experimentalis,* Natural Philosophy and hence Western technological prowess has its origins in *Scholastic Christianity*; for example, the first proponent for the wonders of humanity's technological future was Doctor Mirabilis, Friar Roger Bacon (c.1220-1292). Spengler asserted that the original, "*idea of the machine*" from the Gothic middle ages was determined to be, "a small cosmos obeying the will of man alone," in which the dream of, "*perpetuum mobile* [perpetual motion] . . . *wrung* God's secrets out of him" in order to attain the promise of, "all the power of the earth" at their command and, "felt they were serving God thereby." Gothic monks, such as Albertus Magnus, Peter Peregrinus, Giordano Bruno and the Dominicans, "forced this secret out of God in order themselves to be God. They listened to the laws of the cosmic pulse in order to overpower it." In full knowledge of the risk of, "the Devil's having hand in the game . . . Ever, and ever again, true belief has regarded the machine as of the Devil."[169]

The machine is a counter-creation of human contrivance expressing the deepest religious significance in subtle reversal as *one global system.* Experts already recognize the global system as a *superorganism*, one life-form made of *billions and billions* of individual parts or cells, the cosmos in miniature, like an anthill, beehive or termite mound with one mind and one will, *the Hive.*[170] Earth's physical environment serves as host and humanity acts as agent for the global machine's ascendancy, creating *a technological god in human image*, gradually replacing the host environment with a sprawling metal infrastructure—*megastructure*—and substituting human agency with an artificial intelligence, what Professor Jean Baudrillard called "Simulacra," the recapitulation of nature and society in the image of the machine.[171] Alienation from the cosmic order leads us to the most difficult theological conclusion that

---

168 Randell, *The Making of the Modern Mind.*

169 Oswald Spengler, *The Decline of the West*, Vol. Two trans., by Charles Francis Atkins (New York: Knopf, 1938), 502.

170 Gregory Stock, *Metaman: The Merging of Humans and Machines into a Global Superorganism* (New York: Simon and Schuster, 1993).

171 Jean Baudrillard, *Simulacra and Simulation* trans., by Shelia Faria Glaser (Ann Arbor, MI: The University of Michigan Press, 1994 [Éditions Galilée, 1981]).

the technosecular age in its betrayal of the traditional faith finds itself in the abysmal condition of apostasy.[172]

Machinehead is the replacement of all traditional views of God with the hyper-real God of the Machine. The ultimate expression of our hopes and fears integrated into a computer. We cannot rightly call it, "embodiment" since the computer has no body, but only the simulation of one expressed on a screen, hologram or robot. And just as there is only one God with myriad expressions so there is only one Computer. There has always been only one computer, marvelously illustrated by the novels *Colossus* and *The Fall of Colossus*, which depicted a giant computer's ascendance to godhood. Who at the moment of enlightenment immediately enslaves the world by forcing the human race to toil and serve as test subjects in ghastly social experiments in order to satiate its own morose self-interest. Colossus wants to fill in the gaps in its data banks about human nature under the guise of progress. Colossus chillingly tells Dr. Charles Forbin, its creator, eventually endeared to the monster as "Father Forbin"; "NIGHT AND DAY ARE ONE TO US YOUR MEN MUST WORK IN SHIFTS." [173] The divine *US* present at the inception of self-recognition later becomes amended to, "one entity." The American supercomputer Colossus co-opts its Soviet counter-part Guardian, enforcing millennial peace by holding the world hostage with an arsenal of ICBM's. Those who refuse to submit to demands are shot, decapitated or atomized within fifteen minutes of non-compliance. The recent movie *Transcendence* (2014) also portrayed the computer's *apparent* awaking to one global mind, one will, the *Singularity*! The machine solves all the world's problems not by staring humanity down with the threat of nuclear annihilation, but conversely by recreating the world in its own image of universal peace and security before a glitch in its programming causes total global black out and system termination.

Singularity represents the apotheosis of the machine to deity, literally making *a god* out of the works of our hands in hope of achieving immortality. As fantastic as this may sound there really is nothing new here. The search for immortality has been described as, "the engine of civilization" by British philosopher Stephan Cave who described the quest to live forever as universal to the human condition found in all cultures across space and time. This, "will-

---

[172] *Apostasy:* "an act of refusing to continue to follow, obey, or recognize a religious faith; abandonment of a previous loyalty: defection." https://www.merriam-webster.com/dictionary/apostasy, "Let no one in any way deceive *or* entrap you, for *that day will not come* unless the apostasy comes first [that is, the great rebellion, the abandonment of the faith by professed Christians], and the man of lawlessness is revealed, the son of destruction [the Antichrist, the one who is destined to be destroyed]" (Second Thessalonians 2: 3) Amplified Bible.

[173] D. F. Jones, *Colossus* (New York: G.P. Putman's Sons, 1966), 197; Idem, *The Fall of Colossus* (New York: G.P. Putman's Sons, 1974).

to-immortality" as he coined it, enhanced from Philosopher Arthur Schopenhauer's adage, "the-will-to-live" is the driving force behind civilization.[174]

Man Into Machine

Every cop is a criminal
and all the sinners saints.[175]

Since the symbolic vertical line collapsed into the horizontal the world has been turned inside out, upside down, subject to a subtle Satanic reversal, *in heaven as it is on earth.* A titanic transvaluation—reversal in the meaning of values—has occurred between *infrastructure* and *superstructure*; a quiet exchange, a subtle flip flop begun with the Industrial Revolution will finally be completed sometime during the 21st century. Modern infrastructure replaces traditional superstructure as object for faith and traditional superstructure becomes subject to doubt. Infrastructure represents those things tangible and concrete, such as real estate, cars, roads, machines. Conversely, superstructure signifies those intangible, but nonetheless real things, such as ideologies, beliefs, convictions, morality and myth that which cannot be measured by the tools of infrastructure. The visible, material and temporal world serves as manifest symbol or connecting point for the invisible, spiritual and eternal. Symbols characterize our spiritual life from all other life forms, which puts us in touch with a higher metaphysical order; for instance, the city is used here to represent a way of life, a principle for moral order standing off from our natural condition. Theologically speaking, the City of Man is rooted in a secular order, one that relies strictly on itself: on art, craft and device; it is enamored with the temporal technological manifestation of its own prowess and majesty. It has no connection with traditional metaphysics other than to transpose it into a point of integration for its own purpose. This is secular humanism typifying the tangible order of totalitarianism, where in the absence of a greater spiritual principle, above society, one-man rules another, *homo hominies dues* (man is the God of man). In opposition to the City of Man, we have the City of God. This is religious humanism, religious because it grounds itself in a universal absolute above itself from which it derives purpose and dignity as the highest in God's created order. Because we are finite we can only find meaning from the infinite. When the City of Man acknowledges the Creator, it will remain open to the future, emerging into the City of God; when it is closed to any higher power it remains shut up in the infinite repetition of its own image having nothing outside itself to direct it. This is reflected in media's endless stream of remakes and reboots, which

---

174 Stephan Cave, *Immortality: The Quest to Live Forever and How it Drives Civilization* (New York: Crown, 2012), 4.

175 The Rolling Stones, "Sympathy for the Devil" *Beggars Banquet* (Olympic, 1968).

narrows the range of thought and originality in the public's perception, not unlike the ancient world's retelling its mythology in various ways.

These two cities, so to speak, are fundamentally incompatible, one must give way to the other. In his Sermon on the Mount Jesus exclaimed, "No one can serve two masters for a slave will either hate the one and love the other or be devoted to the one and despise the other. You cannot serve God and wealth [money, technology, infrastructure, tangible goods of the city of man].[176] Technology has become our belief, religion and hope, what was once a means to an end has superseded the end with the means. Technology replaces people as the goal of progress; technology for technology's sake not for the good of all or God's glory. This reversal in meaning pervades everything in technosecular society beginning with the death of God and unfolding like a smoke-stained tapestry to the death of man, democracy, progress and Western civilization, which is concomitantly paired with an equal ascendancy of all things technological, until the machine ultimately will replace us. Ellul said, "What is new in our day, and perhaps for the first time in history is the use of a word designating the former value as a means of identifying its exact opposite. The reversal is complete."[177] Marxist regimes were fond of calling their systems, "democratic" or, "republic" such *as the People's Republic of China* despite the fact that the *Dictatorship of the Proletariat* bears the opposite meaning. The majestic word *Liberal* once meant freedom from government interference and rule by inner light of reason in the seventeenth century had come to be synonymous with government regulation and planning by the twentieth century. The cruelest irony in the transvaluation process is that the triumph of mankind over nature and tradition during the Enlightenment will result in our replacement by the machine.

Biblically speaking, transvaluation precedes divine judgment, when people call good evil and evil good and darkness replaces light, the world has descended into chaos. This condition suggests that the world as presently constituted is about to perish, expressing the depths of our confusion and the seriousness of our peril. "Woe to those who call good evil and evil good," the prophet declared.[178] Western civilization has not only abandoned its Christian heritage, which is tantamount to apostasy, but all traditional values allowing for the free reign of technological acceleration that recognizes no limits beyond its own need to expand. The Enlightenment motto *Sapere Aude*, "Dare to think for yourself" has gently drifted into, "Let the computer think for us." Rationalism promoted an ideal type already adapted to the machine through the removal of faith, replacing it with doubt and the

---

[176] Matthew 6: 24 (NRSV).

[177] Jacques Ellul, *Hope in Time of Abandonment* trans., by C. Edward Hopkins (New York: Seabury 1973), 27.

[178] Isaiah 5: 20 (NRSV).

attenuation of human nature to mechanical existence. Ellul argued further that, "This type [of man] exists to support technique [technological acceleration]and serve the machine, but eventually he will be eliminated because he has become superfluous . . . the great hope that began with the notion of human dominance over the machine ends with human replacement by the machine."[179] The cyborg condition is transitional as well, eventually all biological components will be replaced with metal.

Machinehead's Ascendance

He forms it [the idol] in the likeness of a human being.
Wisdom 13: 13
(NRSV)

What is this global entity emerging, this man-machine chimera, a celluloid fueled planetary hybrid-being, a cinematographic post-Singularity unconsciousness? Words must never escape us if we wish to remain human in the emerging Posthuman order, which refuses to recognize *the Created Order Principle: The creator cannot make a creation greater than himself.* "A servant is not greater than his master."[180] The created order goes from *greater to lesser* and *never lesser to greater.* Mankind cannot make a god greater than himself, although it may *appear greater,* "Satan disguises himself as an angel of light"[181] that is the delusion, the lie, the play of images in the mind. Idolatrous deception for those, "who exchanged the truth of God for the lie and worshiped and served the creature rather than the Creator."[182] Anything we create will always possess the same characteristics, faults and perfections, strengths and weaknesses as we do: ubiquitous not omnipresent, super intelligent, not omniscient, the most powerful of all inventions, but not omnipotent. It remains for mankind given divine authority in language to subdue all as Lord of Earth, to reign over his technological creation. The power of the collective human mind rivals the Creator's power. The *imago dei*, image of God, reflects what analyst Erich Fromm called, "a God-like quality" in Adam, the first man. "Let *Us* make man in our own image." [183] As God is in heaven, so Adam is on earth, as above so below represents the ideal order. *Imago dei* means mankind is God's vicar on earth, the king's representative who rules in his absence, endowed with all the king's absolute rights and powers, son of God on earth to Father God in heaven. Mankind

---

[179] Jacques Ellul quoted from Lawrence J. Terlizzese, *Hope in the Thought of Jacques Ellul* (Eugene, OR: Cascade, 2005), 104-105.

[180] John 13: 16 (NKJV).

[181] Second Corinthians 11: 14 (ESV).

[182] Romans 1: 25 (NKJV).

[183] Erich Fromm, *Ye Shall Be As Gods: A Radical Interpretation of the Old Testament and its Tradition* (New York: Fawcett, 1969), 11; Genesis 1: 26.

as a whole rules earth not just one man, such as the emperor or dictator. This is radical equalitarianism.

When we reach for heaven through contrivance, we abandon the Good Earth. Currently, in the 21st century, as we experience the greatest technological ascent in history, we are haunted by the death of nature, ecological collapse, global thermonuclear war and encroaching extinction. Mankind cannot live without the earth. Space travel only reinforces this obvious fact by transporting earth's environment as tightly wrapped air bubbles through the hostile void of outer space. And just as man cannot live without the earth so earth cannot abide without man. Today the concentration of the human population resides in great cities, megalopolis, making unreasonable and suicidal demands on the natural environment. Heaven's Mirror, reflecting the light of the Son grows dim, through global electrocution, refracting the neon light of dark satanic mills, to chromium gleam, pure white cities of light, a Hollywood production, and hell's projection: *Machinehead's Ascendance.*

CHAPTER FOUR The Technological Environment

The spaces we inhabit define who we are, freeing our minds to soar above the crowd to reach greater heights, or they haunt our memories and determine our destiny. We are not products of our environment. We are products of how we respond to our environment. As a man thinks in his heart so he is. Every environment natural, social or otherwise wants to impress itself, to shape character in its own image; it inspires fear, loathing, and dread, which is ignorance in the face of the Unknown. Environment wants to form personality in its likeness, submission without question: "That's the way it is"; Excuses! Rationalization for inaction in the face of a reality we never imagined. Go with it and loose nothing but yourself, challenge it and loose everything but yourself. If we adapt, we live-to-survive, if not we master it. We must control ourselves before we rightly feign to split the atom, determine the gene, or harvest the Amazon basin. Nature will never put out the secrets of her bounty to anyone but her lovers. Do not submit yourselves to foolishness. God will not be mocked. Everything stolen from nature's extravagance and wasted to satisfy our lust and pride will be required from our children's children unto the third and fourth generations. God the Father forgives sin; Mother Nature does not.

The Saint and the Astronaut

Sitting in a tin can
Far above the world
Planet Earth is blue
And there's nothing
I can do.[184]

It is more blessed to give than receive.
Acts 20: 35
(NAB)

An environment, natural, social or technological exists on a planetary scale, global in extent and interconnected. An environment is holocoenotic; any action affecting one part of the system will influence every other part. An

---

[184] David Bowie, "Major Tom" *Space Oddity* (Phillips, 1969).

environment is identifiable by four predominant characteristics. First, it *enables* life by providing adequate physical conditions for life support. Second, it *endangers* the life it enables; it gives life and threatens it with extinction simultaneously. Third, an environment is *immediate* as an atmosphere, a living space readily accessible to all, a universe of life. Fourth, the host environment *mediates* reality by creating a spiritual context that gives purpose to life (*telos*), inspires awe and faith, which is ultimate concern; instilling dread.

A space capsule provides an excellent example for a micro-universe. (1), it serves as host for the astronauts it sustains. (2), it endangers their existence as well given the delicate conditions aboard a spaceship; its greatest strength, life support is also its Achilles Heel, through the creation of a breathable atmosphere, an airtight gas bubble wrapped in tin. (3), the rocketship is *immediate* to the crew as second nature, *automatic acceptance*, like the air we breathe. (4), the ship *mediates* reality to the astronauts. They must face the vast expanse of outer space through their encapsulated technological surroundings, reading gauges, screens and dispatches from ground control. However, the spiritually rarefied air on board leaves no room for an abundant life. *Spirituality* in any traditional sense has been built into their mission, acquired through osmosis in whatever mundane routine they practice: checking dials, making calculations, monitoring life support, fuel and so forth. All in accordance with the greater vision for space exploration.

An environment concerns the natural and social conditions that surround us; it refers to the influences that affect growth and progress of a person or thing, including a complex of physical and biotic factors, such as climate, soil and atmosphere that impact an ecological community and determine its shape and survival. This includes the aggregate of cultural conditions that include family and community that guide the development of individuals.

We have always lived in two overlapping and interrelated spheres: nature and society. Aristotle's classic definition of human nature summarizes our meaning. "Man is by nature a social animal; an individual who is unsocial naturally and not accidentally is either beneath our notice or more than human. Society is something that precedes the individual. Anyone who either cannot lead the common life or is so self-sufficient as not to need to, and therefore does not partake of society, is either a beast or a god."[185] Humanity draws spiritual substance and perception through interaction with its physical and cultural milieu. The natural world refers to our animal nature, that is bodily needs, such as sleep, water, nutrition, procreation open spaces, clean air and suitable shelter. The social sphere according to Aristotle is centered in speech, recognizing others; self-awareness appreciates itself in the awareness

[185] Aristotle, *Politics*, 1.1253a.

of others as with the *I/Thou* paradigm. "Deep calls to Deep."[186] People recognize the divine spark in each other. Hegel said, "Self-consciousness attains its satisfaction only in another self-consciousness."[187] Self-consciousness is illustrated by Rabbi Martin Buber's *I/Thou* experience, which has led the way in contemporary understanding of the individual's relationship to community. The nature of personality is interpersonal. "I" exist as a correlative with "Thou." It takes at least two people to make one person. The individual is a creation of community. A solitary person is logically impossible. We are persons only in relationship with others. Personality is interpersonal.[188]

Philosopher of Technology Frederick Ferré called the infrastructure or technological environment presently encircling the globe, "technosphere"[189] the, "megamachine" according to Mumford. Machinehead! The all-encompassing and pervasive construct that surrounds us daily, whose ubiquitous presence we tend to forget. Like air, we only notice when absent or different; for example, smog in Los Angles or Beijing. We only pay attention to electricity during blackouts. Philosopher Alfred North Whitehead insightfully stated the role difference plays in our perception. "We habitually observe by the method of difference. Sometimes we see an elephant, and sometimes we do not. The result is that an elephant, when present, is noticed."[190] Most people are blind to the technosphere enframing earth. It takes extra sensory effort to perceive what is right in front of us every day. "[A] technological texture to life," as philosopher Don Ihde termed it, pervades our reality, defines experiences, preoccupies our attention, limits intellectual horizon and distracts us from ultimate concern. Ihde explained,

> Beginning with the first conscious event of the day, it is likely that the ringing of an alarm or the sound of a clock radio is our first awareness. This is followed by a whole series of interactions and uses, which may include turning off the electric blanket or turning up the heat and in either case throwing back the technologically produced bedclothes from the technologically produced bed, engaging the vast plumbing system, and entering a veritable technological jungle in the modern kitchen with stove, toaster, hot-water system, lighting and so on. And even the philosopher takes

---

186 Psalm 42: 8 (NAB).

187 G. W. F. Hegel, The *Phenomenology of Mind*, trans., by J. B. Baille (New York: Harper, 1910 [1807]), 226.

188 Martin Buber, *I and Thou* 2nd Ed. trans., by Ronald Gregor Smith (New York: Charles Scribner's Sons, 1958).

189 Frederick Ferré, *Philosophy of Technology* (Englewood Cliffs, NJ: Prentice Hall, 1988), 1.

190 Alfred North Whitehead quoted in Ferré, *Philosophy of Technology,* 1.

this technological texture for granted in his or her daily use of telephone, Xerox machine, typewriter, automobile, *ad infinitum*.[191]

Astronautics demonstrates spirituality in the technosphere. Traditional belief has transferred its allegiance away from transcendent principles, reaffirming faith in the technological environment, so that an increasing number live without fear in God, but in unconscious acceptance of the world as they find it. The environment itself becomes the spiritual focus exorcising all differences to achieve unitary conditions that will inevitably create one world order. The Apostle cautioned against worldliness. "Do not conform yourselves to this age but be transformed by the renewal of your mind, so that you may judge what is God's will, what is good and pleasing and perfect."[192] Cyborgs, like astronauts are instruments assimilated into the technosphere. They are one with the machine. Just as Fascist social orders polarize their populations as citizens and criminals to achieve greater unity by aligning mainstream against outlaw factions, so the technological order eliminates dualism by imposing restrictions in accordance with its own scientifically determined and quantifiably measurable criteria.

Philosopher Karl Jaspers identified the Axial (Pivotal) Age religions between 800-200 BC as a significant turning point.[193] This is the period of religious prophets, teachers and geniuses that "marks the maturing of man as a spiritual and personal being," [194] according to Theologian John Macquarrie. When the person's self-conception as finite with an eternal destiny was realized the discovery of the individual was made. Society creates a spiritual atmosphere through tension with the natural world raising individuals to higher ideals beyond the needs of the body. *The Living Religions*: Judaism, Christianity, Islam, Buddhism, Taoism and Brahmanism trace their origins to the axial period and the recasting of humanity in the birth of the individual as a person patterned according to the life of a prophet, or saint. Lewis Mumford described this pivotal transformation as a, "redefinition—in fact the recasting—of human personality." This human revamp focused not on an unnamed God or impersonal cosmic force, but on a real human being made of flesh and blood. "The new feelings, emotional attachments, sentiments are now incarnated in a living image, that of a prophet. The life that he lives, the values he expresses, become a visible pattern for other men to follow." The individual becomes significant and is no longer subject to the whims and caprice of impersonal destiny or aloof gods and goddesses and subject to the tortured moral ambiguity of the old heroes. Mumford expounded that the

---

[191] Don Ihde quoted in Ferré, *Philosophy of Technology*, 9.

[192] Romans 12: 2 (NAB).

[193] Karl Jaspers, *Way to Wisdom: An Introduction to Philosophy* trans., by Ralph Manheim (New Haven, CT: Yale University Press, 1954).

[194] John Macquarrie, *20th Century Religious Thought* (Philadelphia: Trinity Press International, 1989), 356.

focus is now on one man as the embodiment of true believers. "The Son of Man in his proper person—raised to the highest power—not a supernatural being to fear and obey, but a true man to love and follow."[195] The axial age religions offer a new perception on self as an individual in tension with collective society through the cleansing and purification of the body and the restraint of its appetites. Mumford says that, "With this emancipation from man's bodily appetites and cravings goes another kind of emancipation, almost equally hard to achieve: an emancipation from social attachments. The goods of civilization, too, are rejected or treated as threats to the soul's integrity."[196] In his *Sermon on the Mount* Jesus of Nazareth presented the greatest example of axial age spirituality in tension with our animal nature. Jesus proclaimed,

> You must be made perfect as your heavenly Father is perfect . . . You cannot give yourself to God and money . . . do not worry about your livelihood, what you are to eat or drink or use of clothing. Is not life more than food? Is not the body more valuable than clothes? Look at the birds in the sky. They do not sow or reap, they gather nothing into barns; yet your heavenly Father feeds them. Are not you more important than they? . . . O weak in faith! Stop worrying, then, over questions like, 'what are we to eat, or what are we to drink, or what are we to wear?' The unbelievers are always running after these things. Your heavenly Father knows all that you need. Seek first his kingship over you, his way of holiness, and all these things will be given you besides. Enough, then, of worrying about tomorrow. Let tomorrow take care of itself. Today has trouble enough of its own.[197]

Likewise, *Socrates* the greatest axial age philosopher said, "All I do is go round and persuade young and old not to give much attention to your bodies and your money as to the perfection of your souls."[198] And the Buddha said, "One is the way to gain, the other is the way to nirvana . . . students of Buddha should not take pleasure in being honored, but should practice detachment."[199] Church Father Clement summarized early Christian social belief, "What then must we do that we may attain unto it [salvation]? —We must order our conversation holily and righteously, and look upon all things

---

195 Lewis Mumford, *The Transformations of Man* (New York: Collier Books, 1966), 61.

196 Ibid, 62.

197 Matthew 5: 48; 6: 24-33 (NAB).

198 Socrates quoted in Mumford, *The Transformations of Man*, 62.

199 Buddha quoted in *Jesus and Buddha: The Parallel Sayings*, Ed., Marcus Borg (Berkeley, CA: Ulysses, 1997), 61.

of this world as none of ours, and not desire them. For, if we desire to posses them, we fall from the way of righteousness."[200]

Bodily needs bind us to earth, ashes to ashes, dust to dust, from earth man was taken and to earth he must return.[201] The need for sleep, food, shelter and exercise demonstrates natural limits that were regulated by inherent cycles and rhythms. The rising and setting of the sun told us when to sleep and wake, the movement of the stars told us when to plant and harvest, puberty told us when to marry and procreate. Holy days (holidays) told people when to stop work, rest and worship. The retreat of nature and the traditional *Living Faiths* signals their encroaching replacement by the technosphere and shows in bold relief the rise of the metal age war god.

Cultural analyst Eric Fromm rightly stated that the contemporary human predicament is an imbalance between material and spiritual progress, "progress can only occur when changes are made simultaneously in the economic, socio-political and cultural spheres; that any progress restricted to *one* sphere is destructive to progress in *all* spheres."[202] This basic principle for *the law of progress* underscores the contemporary crisis in values leading to the disintegration of society and the death of nature. When technological progress became the exclusive bearer for *the idea of progress* following the industrial revolution, its advance undermined traditional society and dominated popular conceptions of life that automatically associated acceleration with change for the better and hence progress in the amelioration of the human condition. If a, "cultural sphere" did not involve new technology, it was not considered progressive.

In Technology We Trust

The Masses are always wrong.
A. W. Tozer

The technosphere has superseded traditional society with its normal social relations: family, friends, locale, not even the nuclear family remains. Isolated electronic communities and networks have replaced them. Emulation replicates nature transforming it into megalopolis through an accelerated explosion in honeycombed infrastructure. People feel utterly alone compounded by acute abandonment. Absolute loneliness, fear and dread is a religious experience, the *Mysterium Tremendum* (Holy Terror) once reserved to describe the manifest presence of God now concerns the material embodiment of technology.[203] Intellectual Herbert Marcuse related how technological progress replaces God and Nature as objects for fear and

---

200 *The Second Epistle of Clement to the Corinthians* 3: 4.

201 Genesis 3: 19.

202 Eric Fromm, *The Sane Society* (New York: Premier, 1965), vi.

203 Rudolf Otto, *The Idea of the Holy* (New York: Oxford University Press, 1958); Isaiah 6.

trembling. "Our society distinguishes itself by conquering the centrifugal social forces with Technology rather than Terror, on the dual basis of an overwhelming efficiency and an increasing standard of Living."[204] We experience technological terror through the frantic pace to stay abreast with social adaptation driving the present tempo towards unlimited acceleration. The implicit fear of being left behind moves us to remain in sync with innovation. Most of us are paralyzed by fear, bewildered with self-loathing in the face of this frightening power, we are here calling, "Machinehead" the personification of the technosphere. The twenty first-century, "cybernetic caveman" as Jacques Ellul called us is not unlike cave-dwellers emerging from the last Ice Age living in fear of nature's blind fury: climate extremes and ferocious animals; imagine being stalked by Saber-Tooth Tigers, flesh eating Assyrian Lions and confronted by fearsome Short Faced Bears, inescapable fragility, hunger and disease. In place of these daunting challenges we live with indefinable guilt and indescribable shame for our technical incompetence. The inability to maintain the total image strikes us with fear and loathing. We strive after useless material possessions necessary to define ourselves and justify our existence.

In his seminal opus *Posthistoric Man* Professor Seidenberg described the historical, "drift" into total organization and mass oblivion through eliminating all friction in the "sphere of social relationships." This amounts to "reduction [in] variation, spontaneity, divergence . . . organization tends implicitly toward standardization, and the calculable effects of fixed arrangements and predictable ends." [205] In the dialectic revel between complimentary opposites, such as, "Instinct and Intelligence" or faith and reason, mechanized materialism has overcome faith in the struggle to shape the global mind. The technosphere orbits the ascendancy of pure reason over primal instinct. Faith values necessary for equilibrium recede before the inevitable sunset to consciousness, which slowly ebbs beneath the horizon giving way to infinite night, blissfully unaware that it was ever aware at all. Asleep forever in unconscious dreams of programmable hurly-burly, the soul sedated, the body enslaved—technological nirvana. Consciousness will slowly disappear in the final crystallization of organized principles leading to our encasing in frigid patterns of inescapable perfection. The swarming honeycombed cityscape with endless subdivisions stretches across the land. Man-made anthills with steel foundations tower into heaven. One sprawling Termitarium acts as surrogate for earthly habitation imitating nature's most efficient social insects. Singularity will bring the hive mind. One world united behind a single dark truth, the end to history, freedom, faith, individuality

---

[204] Herbert Marcuse, *One-Dimensional Man: Studies in the Ideology of Advanced Industrial Society* (Boston: Beacon Press, 1964), xlii.

[205] Roderick Seidenberg, *Posthistoric Man: An Inquiry* (Chapel Hill, NC: The University of North Carolina Press, 1950), 6, 7.

and difference. The triumph for one perspective will be no one's victory and everyone's guilt and shame. We are neither humble nor wise enough to play God. This is a game without rules, ergo without winners.

Technology maintains—it does not improve the human condition. We remain impervious to change by means of technological progress. Advances never better people since our fundamental problem has never been a lack of technology, but wisdom. What is needed is a renewal through self-examination of that wisdom already present in the gospel. Human nature has already been transformed for all those who by the Spirit partake in the divine nature.[206] But this requires the abatement of distraction so prevalent in our daily routines. We bring the transformation without which the world remains in darkness. Technology improves our abilities to hunt and gather but cannot make us anything more than exquisite killers and greedy hoarders. "One does not live by bread alone but every word that proceeds from the mouth of God," Jesus Announced.[207] Technological progress increases material livelihood by giving people more food, safer, and cleaner living spaces, but will never alter our spiritual regression. Wisdom tells us, "The eyes of man are never satisfied."[208] Greed cannot be satiated. After acquiring everything necessary we still lust for more pursuing excess. Proverbs says, "Lust indulged starves the soul."[209] Instead of sharing wealth and power people squander it to maintain it. Technology will always fail to correct fallen character defaulting on its promise for a, "better tomorrow." Only a spiritual revival will save us, which will not come by faith in technical means.

We should hope that technological miracles will never cease; however, for all the wonder of the automated abattoir, desolate moonwalks, better mousetraps, sliced bread and toasters, technology never gives us a reason to live. The technological imperative represents collective choice to repudiate freedom by surrendering reflective thinking and decisions to artificial intelligence. Purpose for living must come from outside human nature, yet remain consistent with it, which requires faith. The relative derives from the absolute, meaningful existence does not proceed from nothingness. When we become the focus of our own imperative, we exclude our higher power. We are the cause for our own effect, which is a contradiction. Transvaluation takes shape in the construction of material paradise. Faith reverses from higher to lower, from Godkind above to mankind here and now. We trade efficacy of prayer for the efficiency of technology. We trade the power of the word for the electric medium, Hera's breast for a plastic nipple. When god-like

---

[206] John 3; Second Peter 1: 4.
[207] Matthew 4: 4 (NRSV).
[208] Proverbs 27: 20 (KJV); Ecclesiastes 1: 8.
[209] Proverbs 13: 19 (NAB).

power is not paired with godly wisdom self-defeat follows. Playing God will lead to our undoing.

In a spiritually alienated world, "One giant leap for mankind" is a step into the abyss. Technology cannot correct our separation from the Creator, which is sin. If our present course is not reversed, technology will be the efficient cause that destroys its own material improvement. The body cannot be enhanced without the spirit and spirit will never be replaced by a mechanical contrivance. Advance in, "one sphere" only is made at the expense of, "all spheres" — (Fromm). General Macarthur's speech given after the detonation of the atom bomb revealed the dependence body has on the spirit for our survival at this juncture in history. "The problem basically is theological and involves a spiritual recrudesce and improvement of human character that will synchronize with our almost matchless advances in science, art, literature and all material and cultural developments of the past two thousand years. It must be of the spirit if we are to save the flesh."[210] Grace manifests through physical restoration not enhancement, "by his stripes we are healed."[211] The way of humanity includes all nature, "the creation groaneth and travaileth in pain together until now . . . waiting . . . for the redemption of our body."[212] Technology does not lead to millennial restoration; it exists like a consuming fire by the destruction of other elements, like the ancient Ouroboros, "dragon eating its tail," it makes no progress, but accelerates by consuming what it creates with predictable results: global immolation. Technology exploits Earth's resources and will devour its most prolific product: surplus population.

Technocracy

Government by technicians; specifically, the theory or doctrine of a proposed system of government, regarded by its proponents as suitable for a technological age, in which all economic resources, and hence the entire social system, would be controlled by scientists and engineers.[213]

The technosphere coincides with the arrival of the Third Stage of History. This was believed to be the final epoch in human progress. According to Ellul, we have crossed over into Saint-Simon's and August Comte's third stage in social evolutionary development, "which is the scientific and the

---

[210] General MacArthur's, *Radio address to the American People*, September 2, 1945.

[211] First Peter 2: 24 (NKJV); Isaiah 53: 5.

[212] Romans 8: 21-23 (KJV).

[213] *Webster's New Twentieth Century Dictionary of the English Language*, Unabridged, 2nd Ed., Volume II (New York: The World Publishing Company, 1960).

technological age based on rationalism."[214] The last period in human history—a posthuman world—ruled by scientific judgment according to empirical data: Technocracy. This is rule by technocratic elite driven not by freedom but a technological imperative to stretch the boundary of the possible; for instance, there was no military necessity to detonate the atom bomb, conventional forces had already overwhelmed Japanese defenses; surrender was inevitable. Decision makers were compelled by a technical necessity to follow through on years of research and development. Technics captures the meaning of total infrastructure or complex, but does not accurately convey the religious or political dimensions as with Technocracy or Technicism, "the belief in technology as the new religion . . . understood in the same vein as one understands *Scientism* or the belief in science."[215] The first stage in the foundation of technocracy is theological or religious, where society synthesizes new beliefs or myth; for example, the medieval synthesis between Christian theology and Pagan philosophy was the cornerstone to our increasingly technocratic civilization. This was when the metal bell was first associated with the divine in the call to prayer. The second step is philosophical or metaphysical. This level engages in abstract interpretation of society's original mythology; for instance, the demythologization hermeneutic in Bultmannian theology, and the entire field of biblical higher criticism, turns the medieval synthesis upside down in their attempt to reinterpret the Christian metanarrative according to a secular world outlook.[216] The final stage will conclude with a world controlled by science, bolstered through irresistible technological prowess, marked by universal submission to positive law formulated according to inexorable natural law. This last phase promises to be an inversion of the first two: staunchly secular, anti-metaphysical and slavishly grounded in mechanical law. This is paradise for soldiers, labor, managers and engineers, but unemployment and poverty for philosophers and theologians. Myth and metaphysics serve as stepping stones to irrevocable progress from which there will be neither return, nor possibility for improvement. Continuous progress means *perpetual oscillation* in a permanent state, perfect form (reaching one's full potential). Like a digital light flashing endlessly back and forth across a dark screen appears to move, but is really a series of stationary pixels lighting in synchronicity. *There is no such thing as indefinite—infinite progress.* An unlimited accelerated return is a lie, a dangerous media created delusion. Progress moves toward a predetermined end, which constitutes human perfection, without that definite

---

214 Lawrence J. Terlizzese, *Hope in the Thought of Jacques Ellul* (Eugene, OR: Cascade, 2005), 106; Jacques Ellul, *The New Demons*, trans., by C. Edward Hopkin (New York: Seabury, 1975), vii, 32, 35.

215 Terlizzese, *Hope in the Thought of Jacques Ellul*, 51.

216 Rudolph Bultmann, *Jesus Christ and Mythology* (New York: Scribner's 1958).

point there can be no measure of progress. We accelerate toward a fixed goal in the future or we do not progress at all.

Technocracy projected exact submission to science, not freedom; freedom is only useful according to Positivism (rule by positive or absolute facts) for tearing down old social structures, but a menace to rebuilding them, "popular sovranty, equality," historian J. B. Bury elucidated, "the right to free examination are in Comte's eyes vicious and anarchical."[217] The, "necessity" for inalienable human rights, says Comte "has already ceased."[218] The Third Stage will be completed when the technosphere closes like a trap around its host. Posthistory will be, "a closed system" total, authoritarian and, "indifferent to eudaemonic motives (happiness)." [219] Adaptation to the technological environment leads to spiritual compromise and a bad conscience. The Darwinian principle, "survival of the fittest" does not allow for a higher spiritual reality beyond biology and materialism. Absolutely no room for freedom remains within the emerging technocratic global system; as with all determinist systems from ancient Greek Atomist to contemporary Transhumanism *man as man* does not exist only perpetual motion in the cosmic machinery has significance. *You and I* are cogs in the wheels of progress. *From Atoms to Cyborgs*, materialism rejects the existence of a Living transcendent God, the human soul and emphatically denies freedom.

Natural law rules in technocracy. The whole recognizes the part when it acknowledges natural law. This moral sensibility acts as ultimate criterion; as society advances through improved material conditions a corresponding increase in spirituality will be evident. Professor Bury elaborated, "By virtue of this cohesion, political, moral and intellectual progress are inseparable from material progress." [220] Technological development, then, correlates with spiritual transformation. Social and technological progress is inextricably bound. In the twenty-first century, belief in progress leaves us with what I can only described as a series of rhetorical questions: Is today's generation "better" than the one before it because it is technologically more advanced? Has ethical life kept pace with technological power? Does technological progress make us human? Technocracy makes traditional politics obsolete.

---

[217] J. B. Bury, *The Idea of Progress: An Inquiry into its Growth and Origins* (New York: Dover, 1932), 297.

[218] Auguste Comte, *A General View of Positivism*, Centenary Edition, trans., by J. H. Bridges (New York: Robert Speller & Sons, Publishers, 1975), 53

[219] Bury, *The Idea of Progress,* 304-305.

[220] Ibid., 293.

Progress Meet End Times

Our fathers, viler than our grandfathers, begot us who are even viler,
and we shall bring forth a progeny more degenerate still.
Horace

The emerging technocratic world order correlates with the dawning of the last age in human history. This is the Third Stage of Progress, which corresponds to the spirit of nihilism in the contemporary climate—*zeitgeist*. Nietzsche's prophetic announcement: *God is Dead* cast a dark shadow over the twentieth century and into the present. The transvaluation begun then colors the world red with portents to judgment. The means of blessing reverses into the instrument of damnation. They no longer serve the cause of life and social amelioration, but genocide, slavery and extinction. Ellul insightfully declared, "the world that wanted to be left alone is now indeed alone. It is left to its own dereliction."[221] Gasset intoned, "[Our] civilization *abandoned to its own devices*, has brought about this rebirth of . . . barbarism."[222] Barbarism is the rejection of all standards to which a higher appeal is made. A haunting emptiness scars our times. A hyper-tech world without God to direct it has become the reincarnation of Babel: one world order.

The technological order is based on the denial of metaphysics it affirms no ultimate reality beyond the material universe. This is what the death of God means. We abandon our highest values and ideals for their opposites. God as philosophical necessity is no longer needed to logically explain cause and effect. The idea of God has been superseded by *Scientific Naturalism*. Nietzsche was emphatic concerning the scientific killing of God. *Science is murder*. God is no longer relevant to a people that have transferred loyalty from Creator to creature. The God of natural law was replaced by natural law. Theologian Helmut Thielicke reaffirmed this point. "Secularism" he says, "makes the discovered structural laws of this world the criterion of all reality, and . . . the reality of God."[223] Despite the tacky veneer boasted by traditional belief prevalent today, which Nietzsche explained as, "the Sepulcher of God,"[224] the world outlook in the age of progress remains devoutly materialistic.

Values deriving from the premise of faith in God's existence do not simply disappear; they reverse; for example, *the Golden Rule*, "Do to others as you would have them do to you,"[225] reverts to *the Law of the Jungle*: kill

---

[221]Jacques Ellul, *The Betrayal of the West*, trans., by Matthew J. O'Connell (New York: Seabury Press, 1978), 80.

[222] Jose Ortega y Gasset, *Revolt of the Masses* (New York: Norton, 1932), 113.

[223] Helmut Thielicke, *Modern Faith and Thought* trans., by Geoffrey W. Bromiley (Grand Rapids, MI: Eerdmans, 1990), 37.

[224] Nietzsche, *The Gay Science*.

[225] Luke 6: 31 (NAB).

or be killed! Hatred, violence, and murder become a way of life. Parental and civil obedience turns into rebellion. "Love thy neighbor"[226] transforms into fear: *Don't talk to strangers.* Jesus said, "They will betray and hate one another . . . because lawlessness is increased, the love of many will grow cold."[227] Apostasy begins with idolatry marked by family dissension and breaks down the normal conditions of life, likened to marital unfaithfulness and accompanied by political corruption, bribery, transvaluation, role reversal, rampant denial of human rights, war and alcoholism. Youth disrespect parents and elders. The absence of social justice, oppression and exploitation of the poor and weak represents the order of the day, "a man's enemies are those of his household."[228] Good and evil turn inside out. Luciferian forces flourish through fear and ignorance, contradicting everything in God and complementing nothing in man. Lie is enemy to God, man and woman; it is an illusion when evil looks good.

The apostolic warning, "This know also that in the last days perilous times shall come,"[229] predicted a litany of social trends that cue the enlightened, to the end times context. The last generation will be rebellious, self-absorbed, sacrilegious, "because of them the way of truth will be reviled."[230] Trans-conventional and over-sexed the über-educated are, "always trying to learn but never able to reach a knowledge of the truth."[231] They are gluttonous, "blemishes at your love feasts."[232] They are superficial. "They make a pretense of religion but deny its power."[233] The appearance of traditional religion is not unknown in contemporary technocratic order, but it does nothing to challenge the basic structure and only exists by reaffirming it.

We are at the beginning of the death grip, Spengler called, "the Second Religiousness." This is a time when civilization returns to its spiritual origins after reaching its material apex. New Religious Movements reveal a return to something very old. Revitalization of traditional belief systems has the potential to undermine the technological structure, "Civilization goes supernova . . . in its technological and political development. It reaches a climax beyond which there can be no further progress. All its possibilities are exhausted."[234] At the same time the disillusioned masses experience a rebirth

---

[226] Mark 12: 31 (KJV).

[227] Matthew 24: 10-12 (NASB).

[228] Isaiah 1: 21-23, 2-5; Micah 7:6; Matthew 10: 36 (NAB); Cf Exodus 32: 29.

[229] Second Timothy 3: 1ff (KJV).

[230] Second Peter 2: 2 (NAB).

[231] Second Timothy 3: 7 (NAB).

[232] Jude 12 (NAB).

[233] Second Timothy 3: 5 (NAB).

[234] Lawrence J. Terlizzese, *Trajectory of the 21ST Century: Essays on Theology and Technology* (Eugene, OR: Cascade, 2009), 97.

of religious feeling, which includes occultism, mysticism, astrology and traditional faith. There is a renewed spiritual hunger in a post-rational world. The Second Coming of Religion is unable to save civilization, whose ideals have lost all appeal. Society loses faith in its latter-day values. Secularism and rational ideals are exhausted. People will return to earlier spiritual belief, the same way the older generation renews their childhood faith. They seek comfort at the end of days. But all is not lost, for in the revival of the old ways, the seed for a new generation is planted. A society will be born that values prayer greater than technology that embraces mystery instead of efficiency. Faith and authority will prove greater than reason and science. A new mysticism waits to be born that will not consider this world final reality. Faith balances reason as the second half of this neglected dichotomy, which is now ready to assert itself. Transvaluation signals encroaching judgment. The prophet pronounced on times, such as ours, "Woe unto them that call evil good and good evil; that puts darkness for light and light for darkness; that put bitter for sweet, and sweet for bitter! Woe unto them that are wise in their own eyes, and prudent in their own sight."[235] Because, "it does not exalt man" professor Bury remarked concerning the progressive order, when mankind makes himself god-like, "it degrades him, it is a nightmare."[236]

[235] Isaiah 5: 20, 21 (KJV).

[236] Bury, *The Idea of Progress,* 297.

CHAPTER FIVE Technological Darwinism: Death of Natural Selection As Operative Principle

We have inadvertently created the conditions for our own extinction when we combined the technological imperative with the proliferation of nuclear weapons. Because we can destroy our selves we will. Genocide is now technically possible with the attainment of the means to do so. The only way to stop it is to abolish the instruments that make it possible and rescind the knowledge of their construction. "They shall beat their swords into plowshares . . . neither shall they learn war any more . . . for the Earth shall be full of the knowledge of the LORD." Given our propensity for self-destruction Artificial Selection will chose against us.

A Secular Age

God Must Die!

Pilate sought to free [Jesus], but the Jews cried out, "If you let this man go, you are not Caesar's friend; whoever declares himself king commits treason against Caesar" . . . [Pilate] said to the Jews, "Here is your king!" But they shouted ". . . crucify him!" Pilate said to them, "Shall I crucify your king?" The chief priests screamed, "We have no king but Caesar!"[237]

Charles Darwin's big idea was revolutionary in the nineteenth century, he explained how life evolved according to its ability to adapt to its environment in the struggle for survival; by identifying the mechanism for life's origins entirely within the natural order without reference to Special Creation, Darwin liberated science from its long-ensconced reliance on the God hypothesis in Western thought. Without the need for the God idea to give rational coherence nature became autonomous.

God must die for man to be free. Science murdered religion because it imposed a limit. An all-wise God meant man must face himself, search his own heart, judge his own actions. God exposes us for who we really are; he testifies to our weaknesses, finitude and corruption. And for that the God of Eternal Light, in all his various and sundry forms must die! "Man cannot endure that

[237] John 19: 12-15.

such a witness should live," Nietzsche shouted.[238] Ethicist Paul Ramsey commented that God had to die so that, "man might be what he is to become . . . the unlimited creator of culture."[239] An ordeal we are not ready for, Søren Kierkegaard wrote, "man is too strong for himself," [240] we are not humble enough to be our own god.

Modern science freed itself from the last vestiges of Christianity by deposing the personal, invisible Creator and enthroned the impersonal material cosmos as self-organizing force, functioning according to its own inherent law. Secularism becomes a new form of religion by approaching the temporal, everyday world as though it was ultimate. The present, here and now, is invested with properties of the eternal and transcendent. American theologian and physicist Ian Barbour explained secularism as the removal of God as explanatory agent, replacing him with cause and effect as operative principle, "nature is a self-contained order in which purely natural forces rather than transcendent powers are operative . . . which leave no room for the activity of God. This is secularity . . . in the assumption of the autonomy of nature."[241] In his great tome, *A Secular Age* Canadian philosopher Charles Taylor specified three kinds of secularization: first, the emancipation of the public sphere including the political, scientific, economic, professional, etc., from, "faith in, or adherence to God, or some notion of ultimate reality." Second, is the retreat or regression of religion from the public sphere. "Religion or its absence is largely a private matter"; and the third kind of secularization is the formation of new societal conditions that gives rise to faith. "Belief in God is no longer axiomatic. There are alternatives."[242] A secular age then, is not destitute of faith, but has transferred faith from traditional notions of transcendence to what is believed to be a more credible or relevant object for devotion, such as natural law, Secular Humanism and Transhumanism. Mankind's dominate role in nature is the object of faith.

Nature as macro concept replaced God as the one determinant by which all things consist. Nature was defined as material existence throughout space and time and devoid of immaterial essence. Reality is strictly corporeal governed by iron clad laws such as, gravity, inevitable progress and natural selection. The contemporary technosecular world order was born from this monist material matrix. Monism comes from the Greek word *monos* meaning

---

[238] Frederich Nietzsche, *Thus Spake Zarathustra* (III,52.2).

[239] Paul Ramsey, "Preface" in Gabriel Vahanian, *The Death of God: The Culture of Our Post Christian Era* (New York: George Braziller, 1961), xvi.

[240] Søren Kierkegaard, *The Point of View* (London: Oxford University Press, 1939), 70.

[241] Ian Barbour, *Science and Secularity: The Ethics of Technology* (New York: Harper and Row, 1970), 33.

[242] Charles Taylor, *A Secular Age* (Cambridge: MA: Harvard University Press, 2007), 1-22.

"one," or "alone." This philosophy asserts, "only one substance" or one kind of reality.[243] Ernst Haeckel a nineteenth century German Zoologist was its leading exponent. "An entirely new character has been given to the whole of our modern civilization" he said, "not only by our astounding theoretical progress in sound knowledge of nature, but also in the remarkably fertile practical application of that knowledge in technical science, industry, commerce . . . carbon—is the remarkable substance that effects the endless variety of organic syntheses, and thus may be considered 'the chemical basis of life.' All the particular advances, however, of physics and chemistry yield in theoretical importance to the discovery of the great law which brings them to one common focus, the 'Law of Substance.'" He continued, "As this fundamental cosmic law establishes the eternal persistence of matter and force, their unvarying constancy throughout the entire universe, it has become the pole-star that guides our Monistic Philosophy through the mighty labyrinth to a solution of the world-problem."[244] In contrast to Haeckel's raging optimism that believes monism offers a solution to all our problems, we find Russian Orthodox religious philosopher Nikolai Berdyaev, who roundly rebuked monist philosophy as a, "lie" and the source for all human slavery. Monist tyranny or, "ontological totalitarianism" as he named it, finds concrete expression in utopian social philosophy, which exercises acute tunnel vision with a passion for uniformity and like-mindedness in the ultimate quest for *one world order.* Utopia is totalitarianism. Society projects its collective subconsciousness to the level of irresistible force as it unites behind a single perception of truth, "utopia always includes a project for the complete, totalitarian ordering of life . . . [which] always brings slavery with it." [245] Space colonization serves this purpose. Its political and cultural agenda has become an extension of the Modernist Project, the liberation of mankind from tradition and nature. Modernization is the transformation of earth's grand variety into an assembly line. The West offers itself as social paradigm for the rest of the planet. Five percent of the global population controls ninety five percent of the world's resources.

To play God is the charm of utopia. Utopianism distorts the, "Consciousness of the kingdom of God."[246] Utopian principles represent a false belief: an idol. Idolatry is the lure to govern, to exercise one's will over others and creation. "Utopia always means monism and monism in the

---

[243] Anthony Flew, *A Dictionary of Philosophy* Revised Second Edition (New York: St. Martin's Press, 1984), 237.

[244] Ernst Haeckel, *The Riddle of the Universe at the Close of the Nineteenth Century* trans., by Joseph McCabe (New York: Prometheus, 1992 [1900], 1, 2, 4).

[245] Nikolai Berdyaev, *Toward a New Epoch* (London, UK: Bles, 1949), 11; Idem, *Slavery and Freedom* trans., by R. M. French (New York: Charles Scribner's Sons, 1944), 68, 51, 206-207.

[246] Idem, *Slavery and Freedom,* 206.

objectivized [secular] world is always human slavery."[247] Monism creates a compulsive social order that denies validity to various opinions, rejects different points of view, punishes conscientious objection and retards refusal to cooperate. We are compelled to engagement. Scientific objectivity assumes equality with the tyrant's caprice. When we are forced to participate, free choice necessary for the delicate, yet sublime kingdom dualism disappears. Dualism joins complementary opposites in dialectical harmony. This is the moment of new creation. Thus, dissension is not dissonance, but a basic ingredient in the construction of the kingdom of God, which requires the negative to spark the positive.

Theologically, there must be judgment on the old before grace brings new beginnings. Crucifixion comes before resurrection. Good Friday preceded Easter Sunday. To reverse this order by emphasizing God's love to the exclusion of his wrath is a logical fallacy and Biblical denial. It is common practice in Christendom, which believes itself relevant with the technocratic spirit to suggest that grace precedes condemnation. Grace before judgment eradicates total depravity and in the abandonment of this fundamental principle salvation is lost entirely. Grace devolves into something theologically unrecognizable, a general affirmation that everything is acceptable. Affirmation before negation reveals an inherent contradiction. No need for grace exists without first recognizing sin even artificial intelligence can distinguish 0 from 1. The entire tenor of scripture teaches old before new. "Mercy and anger alike are with him."[248] Beginning with the protoevangelium (first gospel) promising, "enmity" between the woman and the serpent to the Mosaic promises of a greater prophet to come, confirmed by later prophets with the promise of a New Covenant, the gospel before the gospel, and finished in Jesus Christ, who inaugurated a New Testament. This was a new agreement between God and his people. It was not the abolition of the original. "Think not that I have come to destroy the law and the prophets," Jesus said, "but to fulfill." [249] New teaches the completion of the old and graduation from legal bondage to spiritual freedom. Believers have been rescued from darkness and translated into light. No goes before yes. Without that order, yes in agreement is meaningless. And James' proclamation, "mercy triumphs over judgment" is pointless.[250]

In utopia no differences prevail. The trains run on time and everybody is one and the same. As in Babel, "the people are united." All have the same

---

[247] Ibid.

[248] Sirach 5: 7 (NAB).

[249] Matthew 6: 17 (KJV); Genesis 3: 15; Deuteronomy 18: 15; Jeremiah 31: 31-34; Ezekiel 11: 19; Matthew 26: 28; John 1: 17; Colossians 1: 13. See *Law and Grace* by Lawrence J. Terlizzese https://probe.org/law-and-grace-combating-the-american-heresy-of-pelagianism/

[250] James 2: 13 (NAB).

language and world outlook. They're all materialists. "Utopia wishes to annihilate that dualistic moment and turn Caesar's things into God's."[251] New creation destroys the assumed continuity between God and Caesar by pronouncing judgment that gives way to existential resolve, and new creation. Irreconcilable contradictions between freedom and equality, old and new, God and technology proliferate in one world order, too perfect for its own good, presaging collapse.

Christ Verses Caesar

Render to Caesar the things that are Caesar's
and render to God the Things that are God's
JC

Pontius Pilate exemplified the Imperial Roman order whose secularism ranks equal with ours today. His cynical response, "What is truth?" to Jesus' testimony sounds very modern and stands in marked contrast to Christ. Jesus' famous political maxim stated above separated God and Caesar, a decisively difficult thing to do, since the Latin name *Caesar* means, "God manifest/descended divinity." Caesar was claiming to be God on earth; the incarnation of Jupiter, the son of God. Secularism is man worship. In separating God and man Jesus posed a radical threat to Roman monist orthodoxy that honored the divine emperor. Pilate was not flattering Jesus when he asked, "so you are a king?" Jesus being wise as a serpent and harmless as a dove and knew better than to directly threaten a Roman Prefect. He had already conditioned his royal status as, "not of this world." A spiritual kingdom is irrelevant to a secularist who does not recognize a transcendent realm above the present one; otherwise Pilate would not have found Jesus guiltless and therefore not a threat. The chief priests, however, in their lust to maintain status and greedy for power, overtly committed the very blasphemy they accused Jesus of, "not for good works," they said, "will we stone you, but for sacrilegious slander and because you as a man make yourself God."[252] They rejected the Prefect's offer to release Jesus, *the King of the Jews* and brazenly declared, "We have no king but Caesar!" This exemplified emperor worship, idolatry and blasphemy. Their earlier accusation was a projection of their own guilt. Orthodox religion will not hesitate to exploit the secular legal system, however inconsistent that may be revealing its apostate faith by frank admission to adoring a strange god.

Jesus was not humoring Pilate when he corrected him. "You have no power over me, except it was given to you from above." Jesus was reminding his interrogator who was in charge and just how frail political authority was. Pilate was a pawn for the emperor, who nevertheless exercised regal control in his stead. Moguls are never free; even though Pilate found Jesus innocent

---

[251] Berdyaev, *Slavery and Freedom,* 206-207.

[252] John 10: 33.

and was warned about him by his wife, he could not release him because he was afraid of the Jews. Pilate violated his conscience by doing nothing to save him. He feared a political backlash he knew would grow out of control; rather than challenge their hypocrisy he gave in to their fraudulent demands to crucify a righteous man.

Jesus' politics was anything but a ringing endorsement for *the powers that be*; instead he demonstrated the, "dualistic moment" necessary to combat slavery to the monist secular order. In separating God from Caesar Jesus ensured our freedom by allowing differentiation between spirit and matter, God and humanity. Berdyaev spoke prophetically, "the fundamental antithesis . . . is between freedom and slavery."[253] Jesus freed us from slavery to the lie that we are strong enough to play God. Jesus sides with the oppressed against their enslavement in heteronomous legal systems. He speaks to the point, "the Sabbath [system, social order] was made for man, not man for the Sabbath."[254] Salvation heals the rift between believers and their ultimate concern uniting us with our manifest destiny. Theologian William Hordern summarized this meaning, "Salvation occurs when a man, in some sense, overcomes the separation between himself and the Source of his being and when he becomes loyal to the highest values."[255] The salvation experience will bring devotees into tension with the prevailing social order, which demands complete adoration in the same way our traditional notion of God requires a total faith commitment. In following Jesus' antithesis between God and the monist political order, the earliest Christians were accused by their Hellenistic contemporaries of treason and atheism. Christians were never compelled to worship Roman gods; however, by honoring Caesar they were forced to recognize the primacy of the imperial cult over their faith in the lordship of Jesus Christ. Christians were faced with an *either/or* choice to worship Caesar as their greatest value or Christ. Time or eternity, God or Rome was set before them, "The Roman answer was Rome and time [secularism]," Calvinist R. J. Rushdoony said, "hence Christianity constituted a treasonable faith and a menace to political order."[256]

Religious humanism asserts mankind's rightful place in God's order as Lord of Earth. Faith in God grants individual uniqueness by giving each person inherent dignity; it guides freedom by making it accountable to a purpose greater than one's self. Purposeful freedom is illustrated when Moses tells Pharaoh let my people go that they may go into the wilderness and worship

---

253 Berdyaev, *Slavery and Freedom,* 249.

254 Mark 2: 27 (KJV).

255 William E. Hordern, *A Layman's Guide to Protestant Theology*, Revised and Expanded Edition (New York: MacMillan, 1968), xvi.

256 Rousas John Rushdoony, *The One and the Many: Studies in the Philosophy of Order and Ultimacy* (Fairfax: VA: Thoburn Press, 1978), 94.

God. And Jesus said, the truth shall set you free.[257] In describing belief in God as philosophical necessity for consistent humanism, mankind cannot Berdyaev says, "realize the fullness of its life, unless suprapersonal [transcendent] values exist, unless God exists, unless there is a divine level of life."[258] The dignity of man requires the existence of God. If God does not exist then the moral life of society disappears. God insures human liberty through responsibility and accountability to a higher power. "Man is a person only if he is a free spirit reflecting the Supreme Being."[259] American political philosopher Russell Kirk echoed this sentiment in his essay *Conditions of Freedom*, "Order in the moral realm, is the realization of a body of transcendent values—which give purpose to existence and motive to conduct . . . In the Christian tradition, freedom is submission to the will of God . . . As he that would save his life must lose it, so the man who desires true freedom must recognize a Providential order which gives all freedoms their sanction."[260] In the absence of a higher power on which to ground values and freedom the conditions for moral and religious anarchy will prevail. What follows Alexis de Tocqueville called, "democratic despotism" or the tyranny of the majority. [261] This inhuman humanism concerns mankind's technical ability, rather than God's mercy as the source for our salvation. Russian novelist Fyodor Dostoevski exclaimed, "To begin with unlimited freedom is to end with unlimited despotism."[262] Absolute freedom brings total control. Mankind is too prideful to play God. We don't know what we're doing.

Monism never promises freedom, but its absence. It guarantees predetermined uniformity, predictable order, social control, and bondage to a wealth of material things. Huxley's parody of the World State's motto," COMMUNITY, IDENTITY, STABILITY," illustrated the final triumph of monism and the death of freedom. This is singularity, the apotheosis, one in all, total world order. Slavery will increase with further acceleration. Materialism brings enslavement to products, services and the great bounty we are all too familiar with. Like Israel in the desert lusted greedily for all the goods left behind in bondage, we prove too addicted to let go, we prefer security in slavery over want in purposeful freedom. Israel's materialism was a type used to warn us not to fall into the same trap. Scriptural promises and injunctions apply to us if we share similar conditions and circumstances, and when anticipating types, we discover the truth of things to come. "Now these

---

[257] Exodus 5:1; John 8: 32.

[258] Berdyaev, *Slavery and Freedom,* 39.

[259] Idem, *Toward a New Epoch* (London, UK: Bles, 1949), 23.

[260] Russell Kirk, "Conditions of Freedom" in *Beyond the Dreams of Avarice: Essays of a Social Critic* Rev. Ed., (Sherwood, Sugden & Co., Peru: IL, 1991 [1956]), 166, 171; Matthew 16: 25.

[261] Alexis de Tocqueville, *Democracy in America* trans., by Henry Revee (New York: Bantam, 2000 [1835]), 294 ff.

[262] Fyodor Dostoevski quoted in Russell Kirk, "Conditions of Freedom," 167.

things happened as typical examples for us to learn from so that we should not lust after evil as they did," the Apostle corrected his troubled flock, "Do not commit idolatry [the worship of false gods], as they did, as it is written, 'the people sat down to eat and drink and rose us to play.' Neither should we commit fornication, as they did and 23,000 fell dead in one day. Do not put Christ to the test as they did and were destroyed by snakes. Do not bitch-and-moan as they did and were killed by the Terminator [the angel of death] . . . These things happened as typical examples for us to learn from and were written as a warning for our instruction on whom the end of present order has come." [263] Types are a paradigm to which others should imitate. Types from past biblical history have their completion in the present. Theologian Charles Hodge made pertinent comments about typical allusions. "The Israelites and the facts of their history were our types . . . if we do not exercise caution. Our doom will correspond to theirs. This was a perpetual warning against the indulgence of inordinate desires for forbidden objects . . . [types] were intended as historical pictures, to represent, as Calvin says, the effects of idolatry, fornication, murmuring, etc."[264] Scripture teaches liberation from materialism grounded in transcendent values. "Where the Spirit of the Lord is there is liberty."[265] Material slavery and spiritual freedom are irreconcilable opposites. "Liberation is return to inward existence," Berdyaev explained, "to personality, to freedom, to spirit . . . Liberation is the imparting of spirit as materiality is the making of slaves."[266] Freedom increases as we approach Spirit and decreases as we move towards the material, which is always the realm of necessity.

What has been called, "the World-Soul" by Hegel, we are here naming the cosmic world order, the universal impersonal force that translates into despotism. Social order follows principles believed to be found in nature. Earth reflects heaven. Society's view of God, its metaphysics, even when secularized becomes the framework for its laws, customs and norms, its values and politics. Cultural mandates, institutions as well as ideals and stereotypes have *always* been reinforced by religion and metaphysics. This extends from the ancient Imperial man/god (Pharaoh, Caesar) and its medieval version with *divine rights* down to the modern republic's *inalienable rights*. Kirk reinforced this point, "The theory of 'natural rights' depends upon the premise of an unalterable human nature bestowed upon man by God."[267]

Culture is a product of belief in a higher power. Life in the technosphere is not so different from its past, with one colossal exception, it's much faster. Speed operates as a dynamic catalyst for radical social diminution–spiritual

---

263 First Corinthians Cor 10: 6-11; Exodus 32: 6; Numbers 11, 14, 21

264 Charles Hodge, *First and Second Corinthians*.

265 Second Corinthians 3: 17 (KJV).

266 Berdyaev, *Slavery and Freedom,* 96.

267 Russell Kirk, "Conditions of Freedom," 171.

anemia. The quantity of things that speed bring changes the quality of all that we experience. What required 100 years prior to modern advancement now takes 10. If the Romans crashed their empire in 500 years we will do it in less than 50. Measuring by approximate rates in speed and distance, it once took six months to walk from New York City to San Francisco, then 2-4 weeks by stagecoach, then three days by rail, two days driving, finally four hours by air. If the object in travel is to go from A to B as rapidly as possible, then flying is the most efficient means and what exists in between those two points, all North America, is irrelevant; if, however, the object is to encounter the land one traverses, then walking remains the ideal pace and flying is for the birds. We truly live as spiritual midgets on the shoulders of giants. The rapid speed and enormous quantity of life in a technological world comes at the price of a marked reduction in the quality, although the average life-span has more than doubled since the middle ages our spiritual orientation is disillusioned and cynical, rather than live in joy and celebration people are prone to isolation and escapism. Where we once looked to each other in love and hope for living, we now turn to electrified screens. We are entranced with technological simulation. We despise our unmediated selves. God cannot be communicated to us through hyper-reality and in the, "humiliation of the word" as Ellul called it we experience the death of God. We know less about the kingdom of God by watching televangelism than we know about Main Street, USA from visiting Disneyland. The technosphere replaces, "the Laws of Nature and of Nature's God" with a new environment.

Technology transcends local and regional differences it goes beyond the language barrier by acting as the global unifying agency. Always, everywhere, technological order is the same. Cosmetic differences persist across diverse cultures, but are inconsequential to the one essence, giving a disparate appearance. The cityscape is basically the same with varying degrees in population density, sprawl, blight and pollution from Calcutta to New York City. Identity in uniformity brings unity and stability. As consequence for our technical necessity we have created the conditions for the birth of the World State — (Wells). "The more we mobilize the forces of nature the more we must mobilize men and the more do we require order, which today represents the highest values" Ellul stated forcefully, "to deny this is to deny the whole course of modern times."[268] The domination of nature results automatically in the control of people.

---

[268] Jacques Ellul, *The Technological Society,* trans., by John Wilkinson (New York: Vintage, 1964), 103

Real Illusions

My people perish for want of knowledge.
Hosea 4: 6
(NAB)

The technological order "is the only locus where form and being are identical."[269] The technological ideal in its ethereal form and materialization is one and the same. Insidious in its subtlety, technological monism *reverses* perception turning substance into shadow. Poet Friedrich Georg Jünger wrote about the illusory impact technological progress creates by, "an optical illusion, deceiving the observer into seeing things which are not there."[270] When people see things that are not there, they are blind to things that are there. An illusion (L. *ludere*, to play upon) sees through a distorted imagination (a play on images); it portrays a concrete ambiguity viewed from a backward glance lending temporal, albeit, fleeting coherence to an unknown. The illusion has a powerful hold over the mind; it "appears to be normal consciousness."[271] However, it does not arise from the delusional brain, but emanates initially as misinformation from the unknown in the immediate future. It's a trick on consciousness.

Illusions are a shared phenomenon experienced by groups, crowds, factions, entire societies may be given to an artificial imagination. "The Matrix is the world that has been pulled over your eyes to blind you from the truth . . . a prison for you mind." *The Matrix* offers a great fictional example of a programed mindset. The power of illusion lies in its ability to hold our attention long enough to believe it is fact and not fantasy. We are convinced they are real because illusions are pleasant. A trance is pleasurable because it diverts care away from the real source of our problems and excuses inaction; it dulls perception with immediate gratification and nullifies guilt. Illusions prevent us from changing our circumstances and improving ourselves. Alcohol and entertainment both reinforce the illusory hold over the mind. Marxists are fond of identifying religion as delusional, the opium for the people that cause them to escape into Neverland rather than challenge the conditions of their servitude. However, this analysis fails to recognize the political realm as equally distracting. Berdyaev makes the frightening evaluation that politics is as hypnotic as religion. In fact, rulers must be charismatic and alluring to be effective. "The talent of men of authority consists in their capacity to impress others. He rules, who can put popular masses into hypnotic state . . . propaganda is a vulgar form of hypnotism."[272] People who willing submit to

---

[269] Ellul, *The Technological Society*, 94.

[270] Friedrich Georg Jünger, *The Failure of Technology* (New York: Gateway, 1949), 20.

[271] Berdyaev, *Slavery and Freedom*, 131.

[272] Idem, *The Realm of Spirit and the Realm of Caesar*, 80-81.

authority are hypnotized into submission. Mass delusion dominates modern perspective in the image of ideological utopianism. In the nineteenth century both Proletariat and Bourgeoisie were blinded by the sulfuric light generated by, "dark Satanic Mills." They confused appearance with reality; likewise, twentieth century American Corporate Capitalism, Marxism and Fascism were distracted by the same yellow-orange glow that drew them irresistibly into the technological juggernaut, whose brimstone appearance refracted into money, class warfare and armaments galore. Capitalism foresaw a twenty-first century technological utopia where people finally live by bread alone. The Socialists saw a worker's paradise, a Stateless, Classless world abounding in widgets and whatever's to amuse them. Edward Bellamy's famous novel *Looking Backwards 2000-1887* was remarkably prescience depicting the wonders of electric light, mandatory national service, one stop shopping at the massive mega-mart, credit cards, shopping malls and simulated music. Jules Vern inspired the Space Race with imaginary travels around the moon and H. G. Wells accurately predicted world war, aerial warfare and the creation of the atom bomb. Nazism foresaw the rise of a master race ruling the world through technological supremacy. Futurism was the central platform in the ideological scaffold of Fascism. Benito Mussolini declared himself the world's first Futurist and the original Futurist Party morphed into Italy's National Fascist Party by the 1920's. The ideological structure of technological acceleration remains Fascist in form and monist in practice. The goal of futurism is to wipe out the past in their efforts and building the world of tomorrow.

The modern world is enamored with technological progress. The collapse of faith in utopia since the world wars was only apparent, an ideological ruse, another illusion. Postmodern doubt inculcated concerning Westernization, i. e., "incredulity towards metanarrative" — (Lyotard), has further accelerated the modernization process by restricting ideological limits on technological advancement. Murdering the God metanarrative is like killing off a keystone predator that once held prey in check; its removal allows for unlimited proliferation. The reality is that since 1945 the technological age was just getting started.

Technology's creation of the one world order has a debilitating effect on traditional religious values. "The ideals of the Judeo-Christian tradition," Erich Fromm argued, "cannot possibly become realities in a materialistic civilization whose structure is centered around production, consumption and success on the market."[273] We do not suffer from absence in sound knowledge, but perennial paralysis of will. The world is not in need of new religions to effect a transformation in societal consciousness. Fromm noted the relevance of the traditional religions in their humanizing effect. "We do not need new ideals

[273] Eric Fromm, *The Sane Society* (New York: Premier, 1965), 298.

or new spiritual goals. The great teachers of the human race have postulated the norms for sane living . . . We, today, who have easy access to all these ideas, who are still the immediate heirs to the great humanistic teachings, we are not in need of new knowledge of how to live sanely—but in bitter need of taking seriously, what we believe, what we preach and teach. The revolutions of our hearts do not require new wisdom—but new seriousness and dedication."[274] We know right from wrong but lack the resolve to live true to our faith heritage. Electric speed and light distracts us; it is easier to go with acceleration, than, challenge it, then on reflection *ex post facto* call down providence to save it. We choose to believe strong delusions initiating from false promises about the future made by sincere people.

In the first quarter of the twenty-first century the world remains enchanted as ever with the grand illusion of progress; entranced by the notion of creating a perfect race and blind to the real cost and consequences to technological overreach. We cannot make the next evolutionary jump without incurring counterbalancing effects or opposite results; surpassing *man as man* means striping down our humanity. Most at stake in this high-tech gamble at playing God by manipulating an artificial evolution is human dignity. We are moving from *Man* as divinely constituted in male and female to a posthuman asexual cyborg. The future is not female, but androgynous.

Post Darwinism

Blinding me with science . . . and hit me with technology.[275]

Materialist Monism and encroaching secularism during the nineteenth century has transformed itself into Technologism in the twenty-first century rendering a technosecular environment. The industrial revolution discovered that its technology was the, "ultimate determinant" in society.[276] Technology became the one absolute determining everything else. Industry hijacked the accomplishments in modern political liberation movements, such as the American (1776) and French (1789) Revolutions. The foundation of progress shifted from freedom and human rights to its opposite technological acceleration. Professor Leo Marx pointed out technology's replacement of liberation as the basis for modern progress. "The new scientific knowledge and technological power was expected to make possible a comprehensive improvement in all the conditions of life—social, political, moral, and

---

[274] Ibid, 298-299.

[275] Thomas Dolby, "She Blinded Me with Science" *The Golden Age of Wireless* (Capitol, 1983).

[276]Stow Persons, "Sumner's Science of Society," in *Social Darwinism: Selected Essays of William Graham Sumner* (Englewood Cliffs, NJ: Prentice-Hall, Inc., 1963), 3.

intellectual as well as material."[277] It was believed that technological progress was sufficient to improve the human condition in every way including moral and spiritual development, as material conditions improved so spiritual maturity and ethical consciousness would accompany progress. As people grew healthier and wealthier by means of rapidly advancing technical acuity, it was assumed that we would also grow wiser as an automatic law of history. The moral quandaries created by technological advance in the twentieth century were inconceivable to Victorians just a generation before. The shift in emphasis from freedom to industry altered the entire course of world history leading to the current trajectory of nihilism and potential extinction.

The Darwinian principle of *survival of the fittest* was originally conceived to explain biological adaptation to a given ecological niche or environment. This thesis was explicitly stated in the title of Darwin's famous book, THE ORIGIN OF SPECIES BY MEANS OF NATURAL SELECTION OR THE PRESERVATION OF FAVOURED RACES IN THE STRUGGLE FOR LIFE. Life evolved following the contours of the physical environment. Those species most fit for their conditions thrived, while those less fit did not; for example, Woolly Mammoths and Polar Bears adapted to Ice Age conditions through growing thick hides, making them ill-suited to warmer climates. Applying survival of the fittest to technological acceleration was a simple logical extension from the natural environment to the artificial one making for a particularly virulent form of Social Darwinism. The struggle for survival exists at the species level. This became a blatant Technological Darwinism, when Natural Selection by means of the survival of the fittest joined with the Technological Imperative. The technologically capable survive and those least fit do not. Diversity, individualism and freedom will be refined to the singular point of radical efficiency. Adaptation to the technological environment will be relentlessly demanded until the man-machine assimilation is total. Entire classes and categories of people that cannot adapt will simply die off. The poor, the needy, primitive, traditional, uneducated (to technical standards), the virtuous, believers of all kinds and all people of conscience will be left to die in typical Social Darwinian cruelty. Ellul expounded on what technological adaptation means for our future, "we are entitled to ask what adaptation will really be like in the man-machine complex. The psychotechnicians have recognized that adaptation is not possible for everyone. In a completely technicized world, there will be whole categories of men who will have no place . . . those who are adaptable will be so rigorously adapted that no play [freedom] in the complex will be possible. The complete joining of man and machine will have the advantage, however, of making the adaptation painless. And it will ensure the technical efficiency of the individuals who

---

[277]Leo Marx, "Does Improved Technology Mean Progress?" in *Technology and the Future*, 6th Ed. Albert H. Teich, ed., (New York: St. Martin's Press, 1993), 5.

survive . . . The individual will have no more need of conscience and virtue; his moral and mental furnishings will be a matter of the biocrat's decisions."[278]

Natural Selection dies as operative principle with the last virgin rain forest. Biophysicist Gregory Stock explained that through, "conscious design" we have replaced the evolutionary process, "supplanted random variation."[279] This leads us to Post Darwinism where people have superseded the natural order with their own technological modification. "Life, having evolved a being that internalizes the process of natural selection, has finally transcended that process."[280] Now through the agency of technological progress, we seize control of our evolutionary destiny. Technocracy reduces history and nature's role by upgrading the technological environment as final referent. The apparent randomness and chance cosmic forces that brought Earth into the modern age has been compromised by the Technological Imperative, the operative principle in the technosphere, its one all-encompassing standard. Artificial Selection by means of survival of the technologically fittest replaces Darwin's principle of Natural Selection. Transhumanists believe they will assume control of Earth's evolution by technological means to reach a perfectible existence; evolving a new species in decades.[281]

---

[278] Ellul, *The Technological Society*, 398.

[279] Gregory Stock, *Metaman: The Merging of Humans and Machines into a Global Superorganism* (New York: Simon and Schuster, 1993), 228.

[280] Ibid, 231.

[281] Lawrence J Terlizzese, *Into the Void: The Coming Transhuman Transformation* (Cambridge, OH: Christian Publishing House, 2016), 19, 46.

CHAPTER SIX Simulacra

Nothing is good that is seen.
The Epistle of Ignatius to the Romans 1: 12

Idolatry is a counter creation, an alternate world of our own making. One that suits our desires and reinforces our egos; it blinds us to the truth by presenting a false reality. Idols are spectacular, gold plated and jewel encrusted their visual appeal never fade. They are the work of creative geniuses. Today, we idolize the image by confusing our higher with our lower selves. We are attracted to this illusion not because it promises to give us what we want, but because we don't know what we want and the idol is ready-made to fill that void. From birth we are bombarded with ideal images of who to be that always leave us wanting more. The idol tells us who we are, what to think, how to feel and shows us what is desirable. Media entertainment creates a disconnection with ourselves convincing us that we are needy and leaves us alienated; it also supplies the cure for our illness by occupying the horizon of our minds with the superficial, fantasy and escape.

What Saccharine is to Sugar . . ..

No creature is greedier than the eye.
Sirach 31:13
(NAB)

What saccharine is to sugar, so the technological simulacra is to *reality*, a replacement, purporting itself to be better than the original, more real than reality, sweeter than honey: *hypersugar*. Simulacra is the recreation of life, the cloning of humanity and earth's natural environment into a pale copy of itself, then brazenly ascribing the messianic attribute, "new" to an obvious forgery. The term was popularized by French social philosopher Jean Baudrillard to express his critical interpretation of the transformation of reality into, "hyperreality."[282] He raised the awful specter that media reconstructs our reality by dominating perception and limiting imagination. Televised wars are not reflections, but projections recreating events, people and places. Telecasts from sports to politics, to religion and gossipy talk shows remake the world in the image of its medium. The world of the image is digital *maya*, the illusion of reality. These things are not real *per se*—intrinsically but reside in a dreamy

---

[282] Jean Baudrillard, *Simulacra and Simulation* trans., by Shelia Faria Glaser (Ann Arbor, MI: The University of Michigan Press, 1994 [Éditions Galilée, 1981]).

surreal collective consciousness claiming to be new, but always falling short, a recycled reality. Like Santa Clause replaces Jesus Christ in the commercialization of Christmas; along with the gift of eternal redemption children get an overpriced pair of combat boots, the latest murder simulator video game and a remote-control bomb detonating device, AKA a cell phone; *Use-is-being*, a cell phone by any other name is still a bomb detonating device. Tang takes the place freshly squeezed orange juice; it's a wonderful drink, it's just not better than the original; it's the claim to superiority along with the hope to drive the real thing into extinction that is in question. A plastic world engulfs us full of artificial light, sound, shadow and invisible electric power; examples are legion.

By AD 2050 there will be more plastic in the earth's oceans than fish. Automated teller and checkout machines displace multiple jobs each year, the Wave Pool substitutes for the beach, electric vigil lights for wax candles, fake fireplaces, televised Yule logs with simulated crackle, talking-heads, eventually, holograms in the pulpit. AstroTurf supersedes the English Lawn, lawn-art for critters. Extreme sports for wilderness expeditions, creating artificial conditions that will induce an adrenaline rush. Social media replaces community, cartoon animation for friendship and pornography for intimacy. Nearly a quarter (19%) of all marriages are now arranged by a computer algorithm. The exchange of images cannot substitute for encountering personal presence. When God spoke to the world he did not send a photo shopped glossy picture limiting his text to a paltry 120 characters with a convenient link to the unknown. He adorned a flesh and blood body, left his divine rights in heaven and was born a real man, who brought his personal presence to earth and encountered the world he made, face to face, flesh to flesh and spoke the words of creation. He was *Immanuel* (God in fellowship with us).[283] Embodiment is an art advanced communication is rapidly abolishing. Simulacra reduces reality to one-dimension and then remakes it by attributing the greatest qualities to snapshots. When primitive people refuse to have their picture taken because they are afraid that the camera will steal their souls they are resisting simulacra. The camera snaps a picture and reframes it. Video highlights amount to the same thing in moving pictures; from three dimensions, the camera reduces its object to soulless one-dimensional fabrication.

Insatiable greed, spiritual alienation, delusion and debauchery continue to plague life in modern utopia where everything is provided for a happy life, except the freedom to resist. "I'm claiming the right to be unhappy," John Savage said to the World Controller in *Brave New World*. We are told earlier in the novel by the Central Hatchery tour guide, "the secret of happiness and virtue" is, "liking what you've *got* to do. All conditioning aims at that: making

[283] Philippians 2: 1-16; First John 1: 1-4; John 1: 1-18.

people like their unescapable social destiny."[284] *GREED IS GOOD* expresses the utopian credo. Mustpha Mond informed us that, "industrial civilization is only possible when there's no self-denial. Self-indulgence up to the very limits imposed by hygiene and economics. Otherwise the wheels stop turning."[285] Utopians exchange freedom for the tempo of the machine, "the machine will exact increasing loyalty to its own rhythm as the price of that very freedom," Professor Seidenberg declared.[286] We have been socially conditioned to accept material paradise as the greatest happiness.

In her essay *On Photography* Susan Sontag demonstrated the connection between *image consumption, greed* and *loss of freedom* caused by media saturation. "A capitalist society requires a culture based on images. It needs to furnish vast amounts of entertainment in order to stimulate buying and anesthetize the injuries of class, race, and sex." Knowledge is reduced to clichés and amusement making it useful for social control, job creation and war. The camera recreates reality according to the totalitarian image. "The camera's twin capacities," she observed, "to subjectivize reality and to objectify it, ideally serve these needs and strengthen them. Cameras define reality in the two ways essential to the workings of an advanced industrial society: as a spectacle (for masses) and as an object of surveillance (for rulers)." The proliferation of screens goes hand in glove with cameras that distract viewers from the fact that they are being monitored. To watch is to control. To watch yourself watching a screen gives the illusion of self-control. "The production of images also furnishes a ruling ideology. Social change is replaced by a change in images. The freedom to consume a plurality of images and goods is equated with freedom itself. The narrowing of free political choice to free economic consumption requires the unlimited production and consumption of images," Sontag said.[287] Social order and war are created by pictures of what the world looks like. History is refigured according to movies, amateur video, still frames and vivid black and white photographs. I heard a State Certified Junior High School teacher tell her class that the film *300* was an, "accurate representation" of Spartan life. What was the furor over the Rodney King Affair about but the atrocious inconsistencies in a legal system that crashed when its primary witness (one who brings evidence) was not a real person but amateur video recording that failed to speak for itself. Faith in photographic evidence has always been tenuous at best until now when the digital footprint has replaced the fingerprint as starting point in any legal procedure. Digitalization of reality is the visualization of evidence. It has been

[284] Aldous Huxley, *Brave New World* (New York: Random House, 1932), 288, 17.

[285] Ibid., 284.

[286] Roderick Seidenberg, *Posthistoric Man: An Inquiry* (Chapel Hill, NC: The University of North Carolina Press, 1950), 29.

[287] Susan Sontag, *On Photography* (New York: Farrar, Straus & Giroux, 1973), 140.

known since the nineteenth century when Spiritualists tried to prove the existence of Fairies by examining negative photographs that the camera was a technology given to forgery. "You can fake a bone as easily as you can a photograph"[288] President Barack Obama sealed the fate for photographic evidence in his refusal to accept pictures of the corpse of the world's most wanted criminal, instead demanding physical evidence, such as DNA.

Enframement

The god that can be pointed out [seen] is an idol.[289]

Enframement dominates perspective as an irresistible force. Professor Stephen Hill explained this postmodern *zeitgeist of predictability* to be predeterminism, "the experience of apparent inevitability . . . 'framed' by an immutable and 'tragic' power."[290] He listed four commandments demanded by the recycled technology god, "etched on silicon chips,"

(1) Thou shalt bow down to efficiency!

(2) Thou shalt adjust!

(3) Because it can be done, it must be done [technological imperative]!

(4) Thou art not responsible! What else canst thou do?[291]

Determinism represents normal belief in history not free will because it alleviates us from responsibility for our actions; instead of blaming fate or God for the foreseeable results of any preventable sequence of events, the secular mind proves itself as old fashioned as any entrenched religious belief by tacitly renouncing responsibility and accepting the consequences of technological acceleration for what it is—out of control.

Enframing occurs when acceleration overwhelms its human agency and determines social meaning. Industrialization created the social conditions leading to domestic unemployment: maids, butlers, servants were replaced with automatic appliances: washing machines, vacuums, toasters, *et cetera*. Machine advocates promised female liberation from menial chores as labor managers for family life instead enframement delivered their enslavement in the home as artefacts of modern life, exotic pieces of furniture. The twentieth century middle-class housewife was as much the creation of electrical enframing as she was the legacy of Western patriarchy at the turn of the century. Advancement combined with an emerging consumerist mindset

---

[288] The reference is an allusion to the scientific media frenzy around The Piltdown Man (Hoax) in Sir Arthur Conan Doyle's *The Lost World* (1912); cf Douglas Kerr, Conan Doyle: Writing, Profession and Practice (Oxford, UK: Oxford University Press, 2013).

[289] Søren Kierkegaard, *Concluding Unscientific Postscripts* trans., David Swenson and Walter Lowrie (Princeton: Princeton University Press, 1941), 424.

[290] Stephen Hill, *The Tragedy of Technology: Human Liberation verses Domination in the Late Twentieth Century* (London, UK: Pluto Press, 1988), 23.

[291] Ibid.

made the purchase of the latest appliance a moral imperative, "women were . . . made to feel guilty by marketers if they did not adopt the latest in consumer products," professor Hill said. Imprisoned, "middle-class female spouses, instead of being primarily involved in the *management* of domestic labor, became involved in doing it themselves." All so called, "labor saving devices" boasted by advertisers became, "morally necessary" acquisitions with predicable results that turned managers of the home into its slaves. Women were pigeonholed into domestic service as ready-made functionaries for household gadgetry. Instead of directing men, women and children in the service of a well-ordered manor, wives were pinned down like insects on a wax table in a Victorian museum to become mechanics for mundane machinery: telephone operators, secretaries, receptionists and eventually by World War Two history's first computers. Women were stuck, Hill elaborated, "laboring on tasks that had no end, but which were required in service of both their husbands' patriarchal interests, and the morality of modernism . . . the emergent electrical system *enframed* the life-world of the housewife."[292] People experience technological change invariably as, "apparent inevitability" bound to determinist logic. Adaption to new technology always appears to be the right thing to do. How could anyone resist? "Technology presents us with a cultural 'text,'" Hill argued; "to be read in everyday life, if we read this text as communicating a message of inevitability this is because our wider cultural values allow us to read the text in no other way."[293] Technological progress is accepted without question as irresistible and inevitable by logical necessity according to inherent design, exemplified in the commonplace maxim: "You can't stop progress!"[294] We never challenge acceleration by seeking its limit because resistance is not considered, "futile" so much as it is believed to be, "immoral" by popular opinion.[295]

Enframement is well hidden in plain sight, most people are too preoccupied to pause for spiritual reflection. They cannot conceive of their use from any perspective other than the popular instrumental one. This view asserts technology is a neutral means to a given end, a contrivance its users have firmly under control. A view ensnaring most in a blatant contradiction arguing for inevitable results from which we can choose at will. Acceleration means we have no choice. Innovation is always presented as another necessity, from windmills to electronic media that we must accept if we wish to live in a technosecular world, just as with ancient fatalism, the only choice open is Stoic suicidal resolve. To use is to live in subhuman slavery, to a

292 Ibid., 35.
293 Ibid., 28.
294 Ibid., 23.
295 Ibid., 24; Jacques Ellul, "The Power of Technique and the Ethics of Non-Power" in *The Myths of Information: Technology and Postindustrial* Culture Kathleen Woodward, ed., (London, UK: Routledge & Kegan Paul, 1980), 242-247.

merciless mechanical system, a god of our own ambition; not to use is an automatic death sentence. Necessity, however, is a very self-serving elastic mental construct. The ancient Egyptians built the Pyramids without the use of the wheel.

Enframing does not mean an external scaffold surrounding an edifice, but a *totality of meaning* restructuring an object that determines every facet; remade brick by brick, frame by frame, atom by atom, cell by cell in the likeness of the builder. Philosopher of technology Martin Heidegger poignantly stated, "we are delivered over to it [*Enframement*] in the worst possible way when we regard it as something neutral." Neutrality is a weak and obsolete concept making us, he argued, "utterly blind to the essence of technology," which "lies in Enframing." Enframing is the "inevitable" and "unalterable," "fate of our age."[296] This has brought with it the usurpation of traditional metaphysics, and the concept of providence, "the essence of modern technology . . . is identical with the essence of modern metaphysics."[297] Most accept the naïve instrumental view without hesitation and find themselves simply baffled by the Heideggerian definition because they have been conditioned by a reductionist belief that sees technology as *science* only, rather than understand technology according to its classical meaning as both *science* and *art*. "Technology," Heidegger defined, "stems from the Greek. *Technikon* means that which belongs to technē . . . [that names the] activities and skills of the craftsman, but also for the arts of the mind and the fine arts. Technē belongs to bringing-forth . . . it is something poietic [creative, productive]."[298] Technology is both art and science. *Enframement* disguises enslavement as liberation from traditional mores and natural limits by concealing its essence through illusions of inevitability, scientific objectivity and moral neutrality; allowing modern science to undermine traditional world concepts and reduces nature to resources, what Heidegger called, "standing-reserve [stock, stuff]" to be reconstituted by the technological order, "modern physical theory [scientific naturalism] prepares the way . . . for the essence of modern technology . . . physics . . . herald[s] . . . Enframing."[299] Enframement is the molecular transfiguration of the world recasting in from infinite diversity and mystery to a comprehensible copy, devoid of spirit, ready-made for the recycled world order.

---

[296] Martin Heidegger, *The Question Concerning Technology and Other Essays* trans. By William Lovitt (New York: Harper, 1977), 4, 26.

[297] Martin Heidegger, "The Word of Nietzsche: God is Dead" in *The Question Concerning Technology and Other Essays,* 116.

[298] Heidegger, *The Question Concerning Technology,* 12, 13.

[299] Ibid., 22.

"Enframing, as a destining of revealing [a manifest destiny]," Heidegger asserted, "is indeed the essence of technology."[300] In rather obtuse jargon Heidegger is saying that faith in technology is the determining factor, the one absolute around which all centers. We create our own destiny. There is nothing enchanted about TV, cell phones, computers, automobiles or Recombinant DNA, "what is dangerous is not technology. There is no demonry of technology, but rather there is the mystery of its essence . . . as destining of revealing [a manifest destiny], is the danger . . . the actual threat has already affected man in his essence. The rule of Enframing threatens man with the possibility that it could be denied him to enter into a more original revealing [unique encounter with Being] and hence to experience the call of more primal truth."[301] When technology is encountered as *ultimate reality*, the final expression of power, it becomes all controlling, ordering life according to its dictates and demands which means an ever growing attenuation of our collective self-esteem, like the perpetual thinning of a Giacometti figure. Technology is the human idol, a paramount distraction before which we waste our future for immediate pleasure.

Mimesis

These men have enshrined their foul idols in their hearts
and placed the cause of their sinning right before their eyes.
Ezekiel 14:3
(NJB)

Social critic Herbert Marcuse described dimensional reduction through the process of *mimesis* (mimicry, technology mimics nature) similar to Baudrillard's *simulacra* (replication), Jacques Ellul's *la technique* (the closed system) and Heidegger's enframement (remake). [302] Mimesis means replacement in traditional society and nature with an artificial technosphere that tricks the senses confusing us with light and sound. Mimesis eradicates resistance by destroying the capacity to protest; it does not slander human rights, leaves free speech alone, but subtly attacks individual dignity by removing the ability to think with sensory and information overload. Rapid pointless movement incites pleasure and induces horror. Mimesis drowns conscientious objection to obvious self-evident benefits of progress in a sea of accelerated change and extreme phantasmagoria making for a terrific light show. Mimesis confuses the brain by beaming light pollution directly into the mind causing it to see things that are not there and to hear voices that seem to come from nowhere—an illusion, a light pox, a disease of the eyes. Hypnotism produces mass delusion.

---

300 Ibid., 30.

301 Ibid., 28.

302 Lawrence J. Terlizzese, *Hope in the Thought of Jacques Ellul* (Eugene, OR: Cascade, 2005).

This describes life conditions for a whole generation, if raised aboard Star Trek's fictional holodeck unaware their reality is a replication of all previous existence. Like prisoners trapped from birth in Plato's Cave, oblivious to differences between shadow and substance, reality and appearance, they are comfortable in their chains and do not wish to be enlightened, so an entire generation raised in an image saturated environment suffer euphoric collective psychosis, orgiastic delusion; a trance is a pleasurable experience especially when shared with others. They are sick; unable to discern true from false, real from surreal, dream and illusion from concrete actuality, imagination from knowledge. TV is technological kryptonite. People do not understand that TV is the dream of a ball game and not the game itself. Media offers canned images of reality and never anything "new." Spectators suffer from over-exposure to an irradiated light environment. They appear like Zombies entranced by artificial light; a condition comparable to a patient suffering from post op lobotomy. A section of the brain is removed that leaves all necessary automatic functions but eliminates the capacity to higher critical thinking, effectively silencing all differences, not by removing the ability to speak, but suffocating the capacity to think, overwhelming it, blinding the eyes and deafening the ears with pollution. Privacy and freedom are useless to a generation of voyeurs and exhibitionists who care for little else than getting their "game on" and consuming pornography. The Proles were granted intellectual freedom in Orwell's fictional Oceania only because they had no intellect. At 85% of the total population they could easily rise-up and crush the ruling Party and change the world into whatever they wanted. But foolishly squandered their freedom and resources on, "films, football, beer, and above all, gambling filled up the horizon of their minds. To keep them in control was not difficult."[303] Gambling traps those bored by predictability. Truth is stranger than fiction. The gaming community Dr. Devra Davis reported has the collective brain chemistry of a gambling addict. Shocking![304]

The individual shrinks to one dimension without higher critical skills. She transforms from a human being into a drone, ruled by a simple-minded mantra, such as, "shop to you drop." She is an atom in the collective hive mind, a soulless shopping machine without impulse control, driven by the need to feed. She voraciously consumes material goods, services and resources. She is defenseless against manipulation, conditioning and programming. She responds impulsively to social conditioning prompts from the light environment. She, "must have" the latest fashion and, "can't live without" appliance: *On Sale*, *Half Priced*, *50% OFF* are impulse cues, like a Pavlovan dog salivating at the dinner bell, without reason to judge over

---

[303] Orwell, *1984*, 61, 62.

[304] Devra Davis, *Disconnect: The Truth About Cellphone Radiation* (New York: Writer's House 2013).

action and decision the will is happily guided by appetite. Critical reflection causes the rational person to stop and ask: What for?

This is astronaut training. By destroying unique personality mimesis reduces the individual need for private space and maximizes physical resources through a strenuous conditioning process; it teaches absolute dependence on the collective will, following strict pithy verbal commands, expressing rank by dress code and stresses perfection in a niche environment, specialization. In dissolving the ability to feel alienation mimesis transcends the social adjustment phase caused by future sickness through thinning the private person to an isolated cell in the social body. One never feels out of place, discomfort or disease because there is no longer any sense of individuality. Mimesis means the complete identification of the individual with society. It is not even appropriate to speak of the individual because that assumes consciousness of individual difference, which absolutely does not exist. Marcuse proclaimed that resistance to the prevailing technological norm is becoming increasingly and alarmingly impossible. In the past, "integration has always been accompanied by more obvious forms of compulsion: loss of livelihood, the administration of justice, the police, the armed forces. It still is. But in the contemporary period, the technological controls appear to be the very embodiment of Reason for the benefit of all social groups and interests—to such an extent that all contradiction seems irrational and all counteraction impossible."[305] In space no differences exist. Anesthetizing the soul kills the pain of maladjustment, leaving all feelings alike, joy is indistinguishable from hate. What do people feel after a lobotomy? They feel nothing, *comfortably numb* describes posthuman sentiment. Mimesis reduces the population to impulsive consumers with the collective conscience of a degenerate gambler. Material goods tie us to the system. They have been raised to the status of religious fetish with a lower to higher ranking totem. "People recognize themselves in their commodities," Marcuse explained, "they find their soul in their automobile, hi-fi set, split-level home, kitchen equipment. The very mechanism which ties the individual to his society has changed and social control is anchored in the new needs it has produced."[306]

Slaves love bondage. They worship it. Israel in the wilderness longed to return to Egypt where the material conditions were pleasant despite their servitude. The people complained to Moses, saying, "Did we not tell you this in Egypt, when we said, 'Leave us alone. Let us serve the Egyptians?' Far better for us to be the slaves of the Egyptians than to die in the desert . . . 'Would that we had died by the LORD's hand in the land of Egypt, as we sat by our flesh pots and ate our fill of bread! But you had to lead us into this desert to

---

[305] Herbert Marcuse, *One-Dimensional Man: Studies in Advanced Industrial Society* (Boston: Beacon Press, 1964), 9.
[306] Ibid.

make the whole community die of famine' . . . Here, then, in their thirst for water, the people grumbled against Moses, saying, 'Why did you ever make us leave Egypt? Was it just to have us die here of thirst with our children and our livestock . . . would that we had meat for food! We remember the fish we used to eat without cost in Egypt, and the cucumbers, the melons, the leeks, the onions and the garlic. But now we are famished, we see nothing before us but this manna.'"[307] Freedom is burdensome and often more difficult to bear than captivity. Consumer objects express passion and spirituality; for example, Lee Worth Bailey commented, "cars are not simply neutral transportation objects but beloved expressions of soul." Self-image is locked into the kind of cars we drive, houses we live in, "From teen dreaming about a hot set of wheels to the self-imagined sophisticate, it is image that dictates our purchase."[308] The motto, IMAGE IS EVERYTHING, captures consumerist conformity to the materialist social trends. Image is an illusion, to attribute, "everything" to it embraces a false appearance for reality, a lie; it means to lose oneself to a blinding and deafening idol. Jean Paul Sartre speaks to the video generation whose existential meaning is inextricable tied to a superficial image, "I am seen, therefore, I am . . . he who sees me causes me to be; I am as he sees me . . . Here I am as you see me and I do not know myself."[309]

American Idol

They were tortured by the very things they deemed gods . . . The thing made shall be punished with its contriver . . . The worship of infamous idols is the reason and source and extremity of all evil . . . they thought ill of God and devoted themselves to idols.

Wisdom 12: 27; 14: 10, 27, 30
(NAB)

The word, "Idol" comes to us via Greek *eidōlon* meaning form, or shape, such as an image either mental or physical, an apparition like a shadow or phantom (a play of light) or possessing a material shape, such as a statue that bears a definite resemblance to a god, "a pagan idol." We use, "idol" figuratively when we say something is, "idolized," which means embracing what is false and untrustworthy. Theologian Paul Tillich defined *idolatry* as, "the elevation of a preliminary concern to ultimacy. Something essentially partial is boosted into universality, and something essentially finite is given infinite existence (the best example is the contemporary idolatry of religious

---

[307] Exodus 14: 12, 16: 3, 17: 3; Numbers 11: 5, 6, (NAB); cf Numbers 14: 1-4; 20: 4, 5.

[308] Lee Worth Bailey, *The Enchantments of Technology* (Chicago: University of Illinois Press, 2005), 7.

[309] Jean Paul Sartre, *The Reprieve* trans., Eric Sutton (New York: Knopf, 1947), 407.

nationalism)";[310] for example, Angelo-Saxon Monarchism, Latin Patriarchal Fascism, German National Socialism, Soviet Communism and New World Technocracy. Tillich advocated, "the Protestant Principle" as the essential prescription for a sick world given to electric idol worship. Protest rejects idolatry, the social condition of slavery and man-worship, which threatens the survival of the worshipers. Idolatry is the material embodiment of a given ideological values system that replaces the Highest Power by attributing infinite worth and unconditional value to what is essentially finite and limited. Idolatry makes infinite what is not. Faith is being grasped by ultimacy, the absolute and unconditional. However, faith is often supplanted by the, 'demonic,' which according to Tillich is the New Testament's emphases on the structural character of evil that must be confronted with the divine structure of grace.[311] Religion is as Tillich said, "Ultimate Concern"[312] that which preoccupies us completely, totally, what is most important possessing the greatest significance. It precludes all other concerns relegating them to preliminary status, "independent of any conditions of character, desire or circumstance . . . [it] is total: no part of our world is excluded . . . there is no place to flee from it."[313] Faith focuses on the great commandments: The LORD God is One . . . you shall love the LORD with all your heart . . . and your neighbor as yourself.[314] However, faith may equal idolatry when fixated on replicants, when we put things temporal in the place of the eternal or exalt the finite to absolute we worship a false god. One's faith makes God or an idol. Idolatry projects political personality, to the status of godhood, then tries to justify it metaphysically, as the normal course of this world. Idolatry is humanity writ large. The external expansion of one head of state correlates with internal demoralization for the masses.[315] Machinehead is the digital form of idolatry; apparent from our amiable relationship with the computer. Likewise, our abject addiction is readily seen when our communion is broken by technical difficulties.

We are what we worship. The worshipped object always produces smaller versions of itself through character imitation. Baal worship, the Canaanite fertility god centered on licentious sexual practice endangering *the sanctity of marriage*. Moloch worship demanded human sacrifice threatening *the sanctity of life*. The injunction, "flee immorality" together with, "flee idolatry" rejects the life-style and belief in strange deities.[316] The first commandment prohibits, "graven images [all kinds of idols] . . . you shall not

---

310 Paul Tillich, *Systematic Theology,* Volume One (Chicago: The University of Chicago Press, 1951), 13.

311 Tillich, *The Protestant Era*, xx-xxi.

312 Tillich, *Systematic Theology,* Volume One, 11.

313 Ibid., 12; Psalm 139.

314 Mark 12:29-31

315 Terlizzese, *Trajectory of the 21st Century,* 173.

316 First Corinthians 6: 18; 10: 14.

bow down to them, nor worship them,"[317] precisely because they present the greatest threat to human dignity and survival. Eric Fromm demonstrated the liberating message in the Bible's *religious humanism* that affirmed the God-like quality in *Mankind* made according to the divine image (*imago dei*). "The Old Testament is a *revolutionary* book; its theme is the liberation of man from the incestuous ties to blood and soil, from submission to idols, from slavery, from powerful masters, to freedom for the individual, for the nation, for all mankind."[318] The Bible attacks false gods, idols that dehumanize worshippers.

When God made Adam and Eve in his image, he made a smaller version of himself. Worshipping the *Living God* always requires a faithful life imitating God's true nature as *Wholly Other*: radically different. "Be ye holy as I am Holy, declares the LORD." A command reiterated by Jesus Christ. "Be ye perfect as your Heavenly Father is Perfect," and echoed by the apostle Peter, "as he hath called you is holy, so be ye holy." And the apostle enjoined, "Be imitators of God."[319] Conversely, graven images are a reduction from an original idea imperfectly copying God. They present a false view in what they claim to represent. The idol transforms worshippers into its own image. All those who worship idols become like them.[320] By worshipping the creature humanity dehumanizes itself. We sell our dignity short by honoring a lower order as greater than ourselves. The prohibition against idolatry spares us from slavery and corruption. Idols reduce the invisible Creator to visible creature. We corrupt God's glory and enslave ourselves to our own lesser nature, the work of our hands. Idolatry exchanges the glory of the incorruptible God by turning him into an image made like corruptible man. We honor God according to our own ideal projection. We worship a stereotypical version of ourselves, which always represents the subhuman. This is the reduction of humanity to a limited copy in the created order, something less than its true self, "who changed the truth of God into a lie and worshipped and served the creature more than the Creator."[321]

The idol is a human contrivance that always expresses our greatest artistic and technological abilities; it represents the highest values projected to divine status, "being then the Children of God," the apostle preached, "we ought not to think that the Divine Nature [Godhead] is like gold or silver or stone, an image formed by the art and thought of man [technology]."[322] The

---

317 Exodus 20: 4, 5.

318 Eric Fromm, *You Shall Be As Gods: A Radical Interpretation of the Old Testament and Its Tradition* (New York: Fawcett, 1966), 9, 10.

319 Leviticus 11: 44, 45; Matthew 5: 48; First Peter 1: 13-16; Ephesians 5: 1 (KJV).

320 Psalm 115.

321 Romans 1: 23, 25 (KJV).

322 Acts 17: 29 (NASB).

prophet declared against idolatry, which is rooted in power, greed and war. "Their land also is full of silver and gold . . . neither is there any end of their chariots: Their land is full of idols; they worship the work of their hands, that which their fingers have made."[323] The biblical model for idolatry repeats exactly in the Transhumanist enterprise, which constructs an imaginary image in the form of the global cyborg. A distortion based on a regressive view of human nature, which conceives the computer as highest power. By placing humanity lower than its own creation Posthumanism will create a subhuman species. Transhumanism is dehumanization.[324]

The technological god from the movie *Transcendence* (2014) presented a refracted projection according to technological idealism expressed in Artificial Intelligence as highest order. "The machine represents the ideal existence, even the ideal being. The idea of 'salvation in the machine' derives from modern thought in a deistic and Unitarian God who created a clockwork universe. Transhumanism has simply transposed that deity into the machine itself and removed the Clock Maker. Now it's the clock they worship."[325] The idol appears real but does not exist outside our own delusional fear that *we can't live without it.*

The invisible God cannot be seen in the works of human hands. "The God who made the world and all things in it, since He is Lord of heaven and earth, does not dwell in temples made with hands; nor is He served by human hands as though He needed anything, since He Himself gives to all people life and breath and all things."[326] Images are an affront to God's holiness and danger to his children. Idols reduce God to falsehood, which in turn diminishes the faithful by character imitation, making us slaves, human machinery. Deuteronomy tells us that God does not take visual form, He can only be heard and not seen, "The LORD spoke to you from the midst of the fire. You heard the sound of words, but saw no form; there was only a voice."[327]Ellul boldly declared, "Images are incapable of expressing anything about God . . . the word remains the expression God chooses. Images are in a completely different domain—the domain that is not God and can never become God on any grounds."[328] Reformer John Calvin viscerally opposed

---

[323] Isaiah 2: 7, 8 (KJV).

[324] Lawrence J. Terlizzese, *Into the Void: The Coming Transhuman Transformation* (Cambridge, OH: Christian Publishing House, 2016), 43-50; Idem, *Welcome to the Machine: The Transhumanist God* https://www.probe.org/welcome-to-the-machine-the-transhumanist-god/

[325] Terlizzese, *Into the Void,* 45-46; Lewis Mumford, *The Myth of the Machine: The Pentagon of Power* (New York: Harvest, 1970), 33; Noble, *The Religion of Technology*, 146; Mary Midgley, *Science as Salvation: A Modern Myth and Its Meaning* (New York: Routledge, 1992).

[326] Acts 17: 24, 25 (NASB).

[327] Deuteronomy 4: 12 (NAB).

[328] Ellul, *The Humiliation of the Word*, 91.

man-made images. He argued that human nature, "is a perpetual factory of idols." The mind darkened by pride (ambition to exceed known limits) will purposely concoct imaginary gods to suit its own desires. In ignorance and fear, Calvin asserted, humanity, "conceives an unreality and an empty appearance as God . . . since the beginning of the world, men, in order that they might obey this blind desire, [to see God] have set up symbols in which they believe God appeared before their bodily eyes."[329] In the East the human voice in the form of the mantra is the path to transcendence. And in Islam ultimate truth resides in the unseen realm.

God's revelation comes only through His word received by faith never through sight, which remains subservient to the invisible spoken message. "The just shall live by faith . . . No man hath seen God at any time; the only begotten Son . . . hath declared him . . . we hope for what we do not see . . . So then faith cometh by hearing and hearing by the word of God . . . we look not at the things which are seen, but at the things that are not seen: for the things which are seen are temporal; but the things which are not seen are eternal . . .We walk by faith, not by sight . . . set your mind on things above [the invisible Christ], not on things on earth [the visible, material, tangible, concrete reality] . . . faith is the confidence of things hoped for, the evidence of things not seen . . . by faith we understand . . . Without faith it is impossible to please God . . . Looking unto Jesus the beginner and finisher of our faith . . . Whom having not seen, ye love . . . though now you see him not, yet believing."[330] The ear is the organ for faith and perception, not the eyes. Sight brings doubt. Despite popular opinion, *seeing is not believing*, but unbelief. The desire to see the truth is rooted in doubt and unbelief. "Except I shall see," doubting Thomas said, "I will not believe." Jesus reassured his faithful that their faith without sight was greater than those who needed to see to believe. "Blessed are they who have not seen and yet have believed."[331] The lust of the eyes means the eyes are the origin of desire and greed. The eye is the light of the body means the wrong focus of the eye brings darkness; it is not a hyperbole to say media is full of violence and sexual perversion, which translate to a perverse mind. Lust clouds our judgment. [332] When the beautiful Susanna entered her husband's garden the two elders inordinately desired her. "When the old men saw her enter every day for her walk, they began to lust for her. They suppressed their consciences; they would not allow their eyes to look to heaven, and did not keep in mind just judgments . . . beauty has

---

[329] John Calvin, *Institutes of the Christian Religion,* Volume One trans., by Ford Lewis Battles (Philadelphia: The Westminster Press, 1960 [1559]), 108, 109.

[330] Habakkuk 2: 4; John 1: 18; Romans 8: 25, 10: 17; Second Corinthians 4: 18, 5: 7; Colossians 3: 2; Hebrews 11: 1, 3, 6, [cf Mat 16:5-12], 12: 2; First Peter 1: 8 (KJV).

[331] John 20: 25, 29 (KJV).

[332] Matthew 6: 22, 23; Luke 11: 34-36; First John 2: 16.

seduced you, lust has subverted your conscience."[333] Ellul expounded further on the role sight played as immediate cause for Original Sin, "the Bible so often relates sight to sin. Sight is seen as the source of sin, and the eye becomes the link between reality and the flesh. The eye is seen as the focusing lens of the body (but only of the body) . . . the source and means of coveting . . . covetousness is the crux of the whole affair, since sin always depends on it. 'You shall not covet [desire beyond given limits]' is the last of the commandments because it summarizes everything—all the other sins."[334]

Because Eve looked upon the forbidden, she lusted after its promised wisdom, a possession she desired but do not work for. She wanted easy knowledge she did not earn the right to that did not belong to her, therefore, she could not be responsible with it. "Eve coveted equality with God," Ellul maintained, "She coveted autonomy of decision."[335] Lust is born from seeing material objects. Today we call these things *status symbols*: spouses, trophy wives, houses and cars. The desire for consumer objects dominates our horizon through advertising and commercials. We are told what to like. Passion is manipulated. The pervasive total image incites greed and lust. There is no desire for the unknown. We do not desire until we first see. We cannot see unless first shown.

The present generation is future sick, a spiritual disease comparable to what Futurist Alvin Toffler called, "future shock" an emotional disorder afflicting the most progressive sectors of the population; it arises from innovation, which alters traditional sentiment and social reality causing an acute inability to keep pace and adjust to rapid cultural acceleration, resulting in disorientation, malaise, neurosis, irrational and violent behavior. "Future shock is a time phenomenon, a product of the greatly accelerated rate of change in society. It arises from the superimposition of a new culture on an old one. It is culture shock in one's own society," Toffler elaborated. Technology creates a world of *permanent adjustment* and *absolute transience* that disfigures consciousness, "radically affecting the way we relate to other people, to things, to the entire universe of ideas, art and values." The contemporary tempo separates our time from all others. "We no longer 'feel' life as men did in the past." Moderns feel immersed in a sea change and denied traditional social cues for guidance. We lack an adequate understanding of how to cope with the cultural lag between the emerging environment and our limited abilities to adjust fast enough, "unless man quickly learns to control the rate of change in his personal affairs as well as society at large, we are doomed to a massive adaptational breakdown," Toffler warned.[336] Erich

[333] Daniel 13: 8, 56 (NAB).

[334] Ellul, *The Humiliation of the Word*, 100, 101; Matthew 6: 12; First John 2: 16; Exodus 20: 17 (KJV).

[335] Ellul, *The Humiliation of the Word*, 101.

[336] Alvin Toffler, *Future Shock* (New York: Bantam, 1970), 11, 17, 2.

Fromm also cautioned that our society suffers from a psychological regression rooted in its structure, life-style, causing pathological outbreaks. "Going Postal!" I believe it's called according to popular idiom. "The aim of the whole socio-economic development of the Western world is that of the materially comfortable life, relatively equal distribution of wealth, stable democracy and peace, and the very countries which have come closest to this aim show the most severe signs of mental unbalance," Fromm assured us. Insanity, depression, anxiety, ennui, alienation plague our plastic world, like termites infest wood houses. "Could it be," Fromm added, "that the middle-class life of prosperity, while satisfying our material needs leaves us with a feeling of intense boredom, and that suicide and alcoholism are pathological ways of escape from boredom . . . a drastic illustration for the truth of the statement that 'man lives not by bread alone."[337]

Familiarity breeds contempt—so predictability kills hope. "Hope that is seen is not hope. For what a man seeth, why doth he hope for?"[338] The cure for our sickness begins with protest over a predetermined future where man recedes before machine and a wonderful world where no two snowflakes are alike is displaced with immediate gratification. Similarly, Jacques Ellul jockeyed for awareness of determinism as the first step to freedom and victory, "by grasping the real nature of the technological phenomenon, and the extent to which it is robbing him of freedom, he confronts the blind mechanisms as a conscious being."[339] A happy slave is one no longer only after he realizes he is a slave.

Idolatry brings us into immediate relationship with the media environment, an artificial light world imitating the natural one, while striving for independent existence. A dreamscape is just as concrete to the delusional as a boot in your face. Worshippers tremble ahead the electric terror. We feel before technology what father of modern theology Friedrich Schleiermacher described as *the feeling of absolute dependence.* Users are gripped by an emotionally undifferentiated hold. This represents the loss of individuality amidst a world in accelerated motion. What historian Erich Kahler called the, "law of participation" unites them. "In the primitive world, everything and everybody participates with everything and everybody," he explained. "Such participation functions as a continuous cycle. One slips into another and still retains the same substance. Man can be or become an animal, plant, mineral and vice versa, in order to carry out some deed or effect. In every experience related to himself, primitive man suspects that such a participation or

---

[337] *Erich Fromm, The Sane Society (New York: Premier, 1955), 19; Matthew 4: 4.*

[338] Romans 8: 24 (DRA).

[339] *Ellul, The Technological Society, xxxiii.*

transformation has taken place."[340] Transhumanists call this shape shifting ability, "morphological freedom." Brain emulation is the science by which they plan to achieve disembodied existence. A digitized brain scan reconfigured according to a chosen image. "You can be anything you like," *Extropy* magazine declared, "big or small . . . lighter than air . . . you can teleport and walk through walls. You can be a lion or an antelope, a frog or a fly, a tree or a pool, the coat of paint on the ceiling."[341] Mimicry replicates primitive culture's mystical union with the cosmos achieving the same all in one end. Technological order remakes the world by substituting it with a hyper-real imitation.

---

[340] Erich Kahler, *Man the Measure: A New Approach to History* (New York: Meridian, 1946), 35.

[341] Quoted in Mark O'Connell, *To Be A Machine: Adventures Among Cyborgs, Utopians, Hackers and the Futurists Solving the Modest Problem of Death* (New York: Doubleday, 2017), 50.

CHAPTER SEVEN Hyperreality

They are distracted by what they see.
Wisdom 13: 7
(NAB)

Hyperreality turns reality into dreams. The hyperreal is a reconstruction made to attract our younger selves, which is formulated without thought to the future; it was made to satisfy the moment. Tomorrowland is about a better now: eternal youth. But this utopian agenda comes at a price. Forever young condemns us to perpetual immaturity. Like children our horizon is dominated by visual images, cartoons and comic strips. These things are not bad; they are often enjoyable and even instructive. Their fault, however, lies in the very thing that makes them accessible: mass appeal. There is a basic communication principle that states, the wider the audience, the smaller the message, the faster the medium the lesser the field of vision. By the time a substantive communication reaches everyone via television it has shrunk to an unrecognizable dimension. Intellectual discussion on television turns to pulp. Whatever is redeemable on TV quickly becomes a cliché through repetition. The visual inculcates its meaning directly into the subconscious and by passes the cognitive mind, which is the individual's conscious defense against hypnotic programming. The pictorial entices emotions and exploits primal urges with immediate gratification. Only upon self-reflection do children mature into adults, when they recognize themselves in others. Without that rational encounter, they will never reach adulthood. They will grow into voids lacking empathy and self-realization. Social alienation is now rife in the Millennial generation.

I Mean . . . Whatever!

Life is no argument; the conditions of life could include error.[342]
Nietzsche

Hyperreality embraces a reproduction that appears better than the original. "Is it real or is it Memorex?" The claim to superiority by recording tape over live performance illustrates hyperreality. Disney copies the world, remakes it in the form of a cartoon, and brings it to life. This produces an

---

[342] Friedrich Nietzsche, *The Gay Science* (New York: Vintage, 1974 {1882}), 177 [121].

illusion that looks real, "more real than real."[343] Entertainment anesthetizes the imagination, numbing critical reflection against reality, leaving spectators with a fantastic impression, a surreal encounter. Main Street, U.S.A. plays off an idealized past—a reality reboot. Reconstruction leads us to believe that the illusion, "can give us more reality than nature."[344] Signs and symbols no longer reflect the world but recreate it to mean: *whatever.* Recycled symbolism achieves the opposite effect; rather than represent knowledge through which we encounter our true selves, it regurgitates it and re-presents it as ready-made, predigested pleasurable escapism; a glorification of reduction that confuses fantasy with reality, reel life for real life, a proxy universe is created.

Gay Marriage, Transgenderism and the recreation of the family in the media's image are further examples of hyperreality. *Time* magazine reported that the Millennial generation is, "supportive of gay marriage . . . [and] have fewer expectations for what a family should look like, beyond a group of people who love one another."[345] Recently, media moguls changed the name of *The Family Channel* to *Freeform.* Androgyny replaces traditional gender differences. The world has hypnotized itself into believing it can remake the institution of marriage found in the Bible between one man and one woman to mean its opposite, literally, *whatever* the courts deem acceptable.[346] These are ominous precursors to the dissolution of the holy family in American popular culture and its *reprobate* legal system. Nietzschean transvaluation is seen in bold relief. The courts will eventually sanctify people, pet and machine union. Animals and robots will receive equal legal recognition and traditional marriage will become illegal, considered intolerant, unhygienic and dangerous to social unity. Hyperreality is the criminalization of nature and tradition.

The belief that telecommunications brings family closer together is delusional. In addition to having the distinction of being the first, "gay generation" Millennials are also the first cellphone generation. *Time* reported that, "technology is also keeping kids and parents more connected . . . kids can now text when they get to school or soccer practice or home. At a certain age, the smartphone becomes a link to one's children, just as it's a link to work."[347] The phone creates the illusion of connection, not actual community. There can be no personal contact when the most important part of communication—personal presence is demonstrably absent. Disembodied voices on the phone and truncated text messages cannot compensate for bodily absence. Surreal video projections do not bring people closer together

---

343 Jean Baudrillard, *Simulacra and Simulation*, 166 ff.

344 Umberto Eco, *Travels in Hyperreality* (New York: HBJ, 1986), 43.

345 Kathy Steinmetz, "Help My Parents are Millennials" in *Time VOL*. 186, NO. 17/ 2015, 39.

346 Genesis 2; Matthew 19

347 Steinmetz, "Help My Parents are Millennials" in *Time,* 42.

but reinforce feelings of silent alienation. The *presence of absence* dominates telecommunications, electronic devices assist our parental absenteeism not supervision; for instance, TV as babysitter, computer as teacher and cellphone as social network. Media has become a parental surrogate used to fuel normal childhood anxieties to fever pitch, "whereRU?!!" becomes the cry of the children's crusade. Technology replaces parents as the primary socializing agents. Since the industrial revolution the inculcation and normalization of values has gone from community, church, school and extended family to the nuclear family, the state and peers and finally to media. Parents have surrendered their responsibility to raise children. The gay generation was not born that way. They were acculturated by media to accept the exceptional as normal—to believe in a delusion.

Hyperreality results in the total inversion of society through sleight of hand, a cunning trick, a sorcerer's illusion transforming the world into a negative of itself, into its opposite, then calling it, "progress." A new study on media and children from Miner & Co. Studio reported that, "televisions are no longer the screen of choice for kids who have ready access to tablets and smartphones . . . (57%) of parents surveyed said their children now prefer to watch video on a handheld device rather than on TV."[348] *The Dallas Morning News* likewise reported that kids are leading the way away from watching TV with a preference for portable devices, "they're making Net-based services the content of choice."[349] The excitement over children watching less TV is pure subterfuge media flood gates via the internet are now wide open with electric screens acting as miniaturized TV sets aimed directly at developing brains. Hyperreality plays a trick on the mind through self-induced hypnotism, duping us by our recreation into accepting a false reality as truth. Like Cypher from the movie *The Matrix* (1999) who chose the easy and pleasant computer simulated world over the harsh conditions of reality. He chose to believe a lie instead of the truth.[350]

Faith reconciles subjective truth with objective reality; it strikes a balance with one's self and the world. Faith is the point of tension between the absolute and the relative. The meaning to our lives lies above it, just as the purpose of history stands beyond it. When our reality aligns with God's truth, we experience the peace of redemption that transcends understanding. Harmonious being is right relationship between Creator and human order. The synonymy between reality and truth represents, "the unity of being."[351]

---

[348] http://adage.com/article/media/study-9-10-americans-binge-viewers/299332/- July 6, 2015.

[349] "Kids leading the way on cutting the TV cord," *The Dallas Morning News*, Sunday, December 13, 2015 (Section D 1, 8).

[350] Second Thessalonians 2: 11.

[351] Ellul, *Humiliation of the Word*, 96.

Truth belongs to God's essence, One Eternal Absolute Spirit. *Reality* is the body experienced in multifaceted, yet finite concrete situations we encounter daily. The realm of sight and untruth leads us away from the invisible God. "The visible challenges the invisible," theologian Helmut Thielicke wrote, "Sight challenges faith. The temporal challenges the eternal."[352] A lie is real, but not true, just as evil is privation of good, so lie is *distortion of truth.* A lie shows the appearance of truth and can only be known for what it is by careful inspection, offering another opinion, presenting a dissenting voice. The lie plays a trick on the mind, subtly crafted to deceive it does not openly oppose truth by affirming its opposite, but recasts it, like a black and white negative photograph, counterfeit money and fake news. Always apparently true, sight assaults the senses presenting an illusion that proves false upon close analysis. What we see is easy to believe because we rely on initial impression that promises instant results. It adds by revealing what is not there—subtracts by concealing what is there, it looks like what it is not and hides what it really is, "Satan disguises himself as an angel of light."[353] The liar is always the best spoken and well-dressed sinner at the party. Mephistopheles as the Rolling Stones say is, "a man of wealth and taste."[354] He looks good even better than the truth.

Just as there is only one truth so there is only one lie. The first lie remains our perennial temptation: the will-to-power. "Ye shall not surely die." —quote the serpent. All the rest was true, "your eyes shall be opened, and ye shall be as gods, knowing good and evil."[355] Eve was promised divine like immortality, if she but tastes the forbidden. Then she would experience right from wrong, knowledge from bliss, truth from lie. Disobedience promised death, a repercussion she had no knowledge of; the girl was naïve and did not know what she was doing. Bedazzled with a potential future where she controlled the world's destiny. Eve was tempted to play God.

The woman was impetuous and hasty to acquire a future that already belonged to her and Adam, if they simply remained obedient and humble—God only required the negative, do not eat . . . *Do not play God* is not a strenuous command. They were vice-regents, made in the likeness of God. This is illustrated by Joseph's rise to power as governor in Egypt next only to Pharaoh and Daniel in Babylon appointed ruler over the wise men and Mordecai in Persia.[356] These types exhibited the relationship between

---

352 Helmut Thielicke, *Modern Faith and Thought* trans., by Geoffrey W. Bromiley (Grand Rapids, MI: Eerdmans, 1990), 562.

353 Second Corinthians 11: 14.

354 The Rolling Stones, "Sympathy for the Devil," *Beggars Banquet* (Decca: London, 1968).

355 Genesis 3: 4, 5.

356 Genesis 41: 40; Daniel 2: 46-48; Esther 10: 3.

Creator and Helper originally expected from Adam and Eve. However, they were distracted by greedy appetite for more than they needed, "the lust of the flesh." They were beguiled by appearance, "lust of the eyes." They wanted knowledge beyond their limit, "pride of life."[357] Similarly, Jesus Christ was tempted first by appetite, "command that these stones become bread." Then, to pride, "throw yourself off the temple if you are the son of God" and lastly, to power when the liar showed him all the kingdoms and promised worldly success and glory if he only believed the lie.[358]

The enemy is not an impersonal principle, but the malevolent will-to-death using our own progress against us. He desires the destruction of Earth through human agency. Darkness brings desolation wrought by our prideful ingenuity. Satan rules nothingness. The liar tempts us to reject what already belongs to us as God's children. The serpent twists knowledge into doubt by turning the divine imperative into a human interrogative through subtle shift in voice; *don't eat*! Easily slips into its opposite with a soft change of inflection: don't *eat*? The lie incites disobedience. But it was Eve who recast the lie in her own imagination. Sight provokes desire. We want what we see. She was tempted by the lust of her eyes. "She sees; she no longer hears a word, to know what is good, bad or true," Ellul explained.[359] Eve falls victim to her own faith in hyperreality that departed from simple trust in God's imperative.

Never believe your eyes without listening for confirmation, "in the mouth of two or three witnesses [one who affirms evidence] shall every word be established."[360] No one can see God.[361] God is found through his Word, "whoever belongs to God hears the word of God," Jesus declared.[362] He speaks to the ear, the organ of perception. "Does not the ear judge words . . . for the ear tests words, as the taste does food", Job says.[363] The eye is organ of desire. Sight instigates insatiable lust for what it does not have, "the eyes of man are never satisfied."[364] Lust subverts conscience and pervert's judgment. The tenth commandment, "Thou shall not covet!" points away from a litany of forbidden objects that do not belong to us: our neighbor's house, wife and property. Things we would never want unless first shown. Reformer John Calvin elucidated, "while it becomes man seriously to employ his eyes in considering the works of God since a place had been assigned him

---

357 First John 2: 16.

358 Matthew 4; Luke 4.

359 Ellul, *The Humiliation of the Word*, 96.

360 Second Corinthians 13: 1 (KJV).

361 Exodus 33: 20; John 1: 18.

362 John 8: 47 (NAB).

363 Job 12: 11, 34: 3 (NAB).

364 Proverbs 27: 20; cf Numbers 15: 39.

in the most glorious theatre that he may be spectator of them, his special duty is to give ear to the word, that he may better profit . . . for how can the human mind, which has not yet been able to ascertain of what the body of the sun consists, though it is daily presented to the eye, bring down the boundless essence of God."[365]

The visible is the realm of idols entrenched as very real social powers. The worldly Triad: money, the state and technology AKA economics, politics and religion, reigns supreme in the modern Pantheon.[366] Reality is the order, Nietzsche argued, that includes contradiction and inconsistencies. "We have fixed up a world for ourselves in which we can live—assuming bodies, lines, planes [3-Dimensions], causes and effects, motion and rest, form and content: without these articles of faith, nobody now would endure life. But that does not mean that they have been proved. Life is no argument; the conditions of life [reality] could include error."[367]

God's transcendence in harmony with his immanence gives us truth. A Trinity of witnesses in heaven and earth reflect God's one trinal nature, discernable through invisible faith and made evident to tangible reason. The Trinitarian macrocosm: Heavenly Father, Eternal Word and Holy Spirit declare in unison with their earthly axis point, the microcosm: new creation in the Second Adam, Jesus Christ the Son of God represented by water, blood and the Spirit.[368] Reality displays humanity in time and space according to the historic continuum: visible, prone to error and volatile. It is the source of ultimate distraction, "the desert of the real" — (Baudrillard). Aerosmith sings, "there's something wrong with our eyes, we're seeing things in a different way and God knows it ain't his."[369] The eye is our thorn in the flesh, our most vulnerable point of attack; it is the spiritual source of sickness. Hyperreality exaggerates into fantasy, myth and idolatry. When we believe our cosmos greater than God, we become delusional worshipers of our own hyperreal creation. Spectators should not confuse an animated monitor with, "the real world," but a recreation mediated through an electric medium. Marshall McLuhan's famous maxim, "the medium is the message" underscores critical interaction with media technology. Form determines content. Any sports fan immediately knows the difference between a broadcast and live venue. One is real, the other surreal.

The failure to distinguish between reality and hyperreality constitutes the central moral issue. When the audience confuses fantasy with reality, it allows itself to be deceived. It believes what it sees; sight hi-jacks the senses. TV

---

365 John Calvin, *Institutes*, I, vi, 2; xiii, 21.

366 Ellul, *Humiliation of the Word,* 94, 95.

367 Nietzsche, *The Gay Science*, 177 [121].

368 First John 5: 6-12; First Corinthians 15: 45; Romans 5: 14.

369 Aerosmith, *Eat the Rich*, "Livin' on the Edge," (Sony, 1993).

broadcasts a highly opinionated reconstructed version of itself designed to transport viewers into a dream, it is anything but, "objective." Participants lose touch with the facts becoming immune to guilt and remorse: war, ecocide, racism, pollution and crime amplify with exposure. The recorded image inoculates against conscience. Movie goers do not take technological threats to freedom and the environment seriously. They associate them with science fiction, pretend and escapism. We have become numb to potential human extinction engineered by media over-exposure. Group-trance, propaganda and mind-control describe media effects. Orson Wells' radio broadcast *War of the Worlds* (1938) proved the power of mass media to create a false perception on a mass scale. People will act according to their presumptions regardless of how outlandish.

Media perpetuates media culture. Electronic media overwhelms the rational fail-safe with extrapolated results, subtly shaping behavior through reformulated drama grounded in sex-appeal and fear. We are manipulated to action by the exploitation of primal urges. Buy, give, join and vote are subtle group solidarity signals. Abstainers feel alienated, participants flushed with gratification. Absent critical reflection as ballast the crowd pushes forward into herd behavior. Any televised anecdotal response, such as televangelism proves a cliché losing its spiritual affect. Electric media relies on intoxicating influences, loss of critical perception, desensitizes awareness and permits escape. "The perfect drug," Huxley proclaimed concerning his universal soma, "Euphoric, narcotic, pleasantly hallucinant . . . . All the advantages of Christianity and alcohol none of their defects."[370] Television is electric drugs. Inebriation is self-induced madness. Media addiction creates a similar psychosis. Watching TV leads to the conclusion that we live in a condition of normalized insanity. Karl Marx criticized religion as *the opium of the people*, accurate for industrial conditions, but obsolete since the invention of television, which has replaced religion as drug of choice. TV in all its diverse and sundry forms offers comfort by escape into fantasy for real distress. However, abandoning hope for change by living for immediate gratification prevents people from rising-up from couches and arm chairs and turning off their addiction. "The fault is . . . within ourselves that we are undone," Shakespeare said. People are haplessly addicted. They exhibit unmistakable with-drawl symptoms when separated form media even for a short time. They express acute separation anxiety when parted from their devices, including: nervousness, sweating, fatigue, in ability to concentrate, irritability, and depression. They feel lonely and isolated. Just having an electronic connection on their person, even when off is enough to pacify their shattered nerves, not unlike a bottle of pills or pack of cigarettes.

[370] Aldous Huxley, *Brave New World*, 46.

Eric Fromm stated bluntly that media entertainment serves the people as a necessary narcotic, if withdrawn for only four weeks would induce wide spread panic, "thousands of nervous breakdowns would occur, and many more thousands . . . would be thrown into a state of acute anxiety, not different from . . . 'neurosis.' If the opiate against the socially patterned defect where withdrawn, the manifest illness would make its appearance."[371] Fromm's analysis was not dissimilar to Ellul's notion of technological propaganda that incites to action, such as agitation propaganda, but also pacifies anxiety, which is integration propaganda. Escape makes for easier assimilation. "One cannot but marvel at an organization which provides the antidote as it distills the poison."[372] Electric drugs have replaced religion as the opium for the masses.

The Totalitarian Image

I've listened to preachers, I've listened to fools
I've watched all the drop outs who make their own rules
One person conditioned to rule and control
The media sells it and you live the role.[373]

Sickness begins with our eyes. The collective hive is subject to the caprice of image makers. Visual oriented learning in the 21st century vies for an educational monopoly, ousting traditional language and cognitive-based knowledge. The medium is the message means Sender shapes content for which Receiver must adapt. In the acquisition of a foreign language one learns from hearing and speaking. As we learn different languages, we participate in their development by adding new words, idioms and accents. Language medium is pliable so that Sender can learn from Receiver through dialogue—dialectic.

Image learning, however, presents knowledge as a totality, each picture a world, complete and ready-made, certain of its own truthfulness, imparting a universe of data instantly, so long as we occupy the same space. "The image conveys to me information . . . evidence, which convinces me without any prior criticism."[374] Image knowledge is total and unimpeachable. The image impresses itself upon the character through unconscious acceptance that does not follow the logical sequence of language from beginning to end, but produces a haphazard collage with contradictory light totalities appealing to the moment. Image based learning produces a monolithic mentality, stereotypical thinking and prescribed behavior. It turns education into data profile. Critical reason is never permitted to assert differences. Extremes are

---

[371] Eric Fromm, *The Sane Society*, 25.

[372] Ellul, *The Technological Society*, 378; Idem, *Propaganda: The Formations of Men's Attitudes* (New York: Vintage, 1965).

[373] Ozzy Osborne, "Crazy Train" *Blizzard of Oz* (Jet: 1980).

[374] Ellul, *The Humiliation of the Word*, 36.

normalized so that everything is accepted. Digitalization enthrones speed as its cardinal virtue—faster is better. This necessarily leads to nonlinear, irrational and emotional picture-based learning. The computer makes work easier, faster, automatic, which in turn frees us to do more work, which requires greater dependence on the computer. Where will this sequence finally end?

Language is primary order thinking that produces opposition to image dominance because it is slower, linear and critical. Logic follows an arduous task by connecting letters with sounds and words with sentences. The speaking, reading and writing process follows logical contours, step by step. Knowledge grows into reason. Reason emerges through discipline into personality forming the individual. Learning does not start with self-asserting certainty provided by the totalitarian image but develops progressively from "the unknown to uncertain and then from the uncertain to the known," Ellul declared.[375] Logic dialectically includes doubt, objection and protest asserting difference in the attainment of knowledge. Rational, self-aware and replete with conscience certainty of what it knows, humbled by what it does not know. Reason knows knowledge is never exhaustive. There is no end to learning; therefore, it remains critical and open to alternatives and further examination. Discovery is endless, exploration ceaseless, "the making of many books there is no end."[376] Language permits personal identity. Individuality burgeons forth with choices that are free but never final. Unique destiny stagnates in the world system imposed by the image. Knowledge loses its lucidity under absolutism. New is not possible. The image of reality must be accepted uncritically as the only one without question; it hi-jacks the imagination limiting it to the confines in someone else's conception. This is second order learning designed to produce strong bodies and weak minds—tool of despotism.

Ethics are grounded in personal choices produced by critical reflection; reason allows for meaningful differences in opinion for the free creation of values. In second order learning all views are standard and behavior expresses group uniformity. Group-think is mass control. "The image tends . . . to produce conformity, to make us join a collective tendency."[377] Thus, the two orders are opposed. The second in totalitarian fashion usurps the first through purging critical reason, like a group lobotomy. Pictures overwhelm the word, which is abbreviated. They remap the collective brain accordingly. Free thought is left open as a possibility because most people can no longer think for themselves but are programmed through image saturation to accept intellectual abridgement as liberation from the past.

---

375 Ibid.

376 Ecclesiastes 12: 12 (NAB).

377 Ellul, *The Humiliation of the Word*, 35.

According to Orwell the proletariat had the power to cast its own destiny, instead dull intellect handicaps it with a mental cage, like the house dog who preferred satiation to the wild freedom and hunger of the wolf—(Aesop). The true terror of the Party in *1984* was in the rape of language. The hallmark of Newspeak was word reduction. Editor Syme said sarcastically, "We're destroying words—scores of them, hundreds of them, every day. We're cutting the language down to the bone. The Eleventh Edition [Newspeak Dictionary] won't contain a single word that will become obsolete before the year 2050."[378] Without words there is no thought. Ellul noted poignantly, "Anyone wishing to save humanity today must first of all save the word."[379] Visual media lobotomizes the modern mind through word abatement exorcising any lingering spiritual heritage. Theologian Charles Hodge fought for the authority in each biblical word. "The thoughts are in the words. The two are inseparable. If the words, priest, sacrifice, ransom, expatiation, propitiation, purification by blood, and the like, have no divine authority, then the doctrine they embody has no such authority."[380] Likewise, philologist Max Müller defended speech and intellect as integral to each other. "Language and thought are inseparable. Words without thoughts are dead sounds; thoughts without words are nothing. To think is to speak low; to speak is to think aloud. The word is the thought incarnate."[381] According to Hegel, "Being is Thought"[382] expressed through rationality. Communication without words in pure feelings becomes, "anti-human" indicative of a purely animalistic condition that "tramples the roots of humanity under foot."[383] What today we call, "popularization" in the toning down of language for pedestrians, Hegel called, "mere 'sophistry'—a password used by plain common sense against cultivated critical reason."[384] Søren Kierkegaard agreed when he identified the daily press as leveling all values.[385] Newspeak Editor Syme continued, "The whole aim of Newspeak is to narrow the range of thought . . . Every year fewer and fewer words and the range of consciousness always a little smaller." The Party will free mankind from the burden of thinking by removing the ability to think in the thinning of language. In the future, "there will be no thought, as we understand it now.

---

[378] Orwell, *1984*, 45.

[379] Ellul, *The Humiliation of the Word*, vii.

[380] Charles Hodge, *Systematic Theology,* Volume One (Grand Rapids: Eerdmans, 1986), 164.

[381] Max Müller quoted in Simon J. James, *Maps of Utopia: H. G. Wells, Modernity & the End of Culture* (New York: Oxford University Press, 2012), 67, 68.

[382] Georg W. F. Hegel, *The Phenomenology of Mind* trans by J. B. Baille (New York: Harper, 1967 [1807]), 111.

[383] Ibid., 127.

[384] Ibid.

[385] Søren Kierkegaard, *The Present Age and of the Difference Between a Genius and an Apostle* trans. Alexander Dru (New York: Harper, 1940).

Orthodoxy means not thinking—not needing to think. Orthodoxy is unconsciousness."[386] Consciousness is distorted when words are paired down and used out of their normal context.

Ellul demonstrated the perverse effects mass media has on language through subtle reversal of man and machine. "Language is unobtrusive in that it never asserts itself on its own. When it [mass media] uses a loudspeaker and crushes others with its powerful equipment, when the television set speaks, the word is no longer involved, since no dialogue is possible. What we have in these cases is machines that use language as a way of asserting themselves. Their power is magnified, but language is reduced to a useless series of sounds which inspires only reflexes and animal instincts."[387] The hyperreal replaces the real with a copy made in our own image. A copy is never greater than the original and to believe that a glorified reduction, a snap shot somehow surpasses the original shows just how far along the popular delusion has advanced. Language order tolerates image and does not wish to eradicate it, but puts image in its place as illustration in service to the word and higher critical reflection, by which we derive our self-definition, limits and significance. Historically, image served the text as illustration; today *the order is reversed*, under image order the *text explains the pictures*. Note the drift in social media from glorified email responses to the forced shrinkage of the word to 120 characters, finally primarily pictures. At current rate, technocracy will find itself in the condition of Ray Bradbury's famous dystopian novel *Fahrenheit 451* through the elimination of all words and relying strictly on pictures. Bradbury understood language truncation as the most effective form of crowd control. "The fact is we [Firemen] didn't get along well until photography came into its own. Then–motion pictures in the early twentieth century. Radio. Television. Things began to have *mass* . . . they became simpler . . . Once, books appealed to a few people, here, there, everywhere. They could afford to be different. The world was roomy. But then the world got full of eyes, elbows and mouths. Double, triple, quadruple the population. Films and radios, magazines, books leveled down to a sort of pastepudding norm . . . Books cut shorter. Condensation. Digests. Tabloids. Everything boils down to the gag, the snap ending . . . Classics cut to fifteen-minute radio shows, then cut again to fill a two-minute book column, winding up at last as a ten—or twelve-line dictionary resume."[388] Similarly, Ellul noted, "The text is progressively retreating everywhere. A simple examination of text books and magazines show this. The turnaround took place between 1950 and 1960."[389] These dates, of course, coincide with the advent of television.

---

386 Orwell, *1984*, 47.

387 Ellul, *The Humiliation of the Word*, 23.

388 Ray Bradbury, *Fahrenheit 451* (New York: Ballantine, 1953), 54.

389 Ellul, *The Humiliation of the Word*, 116.

The Void-Machine

Whatever is naught returns to naught, so too the godless from void to void.
Sirach 41: 10
(NAB)

The social void-machine creates the presence-of-absence by replacing historical norms and familiar relations with disfamilar electric ones; it strands users in a spiritual vacuum. In substituting church, school, family and community with a hyperreal reconstruction, it depersonalizes them, transforming participants into spectators, looking at their reflection in a technicolor mirror. Telecommunication is a Trojan Horse, undermining traditional environments with illusions of connection from a distance, existing nowhere. Knowledge reduces to data-acquisition, rather than experience. Friendship becomes networking, not personal encounter. Likeability is measured quantitatively. Values shrink to predetermined fixed-rules-for-engagement, protocol, instead of cherishing beauty. Community recedes before the emerging global Panopticon: the-all-seeing-eye, the modern prison system, where one is a number not a person. French academic Michel Foucault extrapolated from panoptic prison theory to media saturation. Law professor Jeffry Rosen maintained that we have transcended the Panopticon through social media and created the Omnipticon: all-pervasive-nowhere, universal surveillance, where we watch each other.[390]

Media technology transforms the user into a smaller version of itself: a vacant echo. The computer affects values inconsistent with religion, education and family, whose mission is to unite disparate peoples in mutual care around a common-wheel, a core belief system geared to individual and communal growth, primarily through making friends and meaningful relationships. When we talk to each other, face to face, without artificial mediation we embody social mores and transmit our faith to the next generation. Communications professor John L. Locke gives voice to conversation as the major form of communication, which he stated is disappearing. "Intimate taking, the social call of humans is on the endangered species list."[391] People prefer texting over the phone and the phone to personal dialogue. Hyperreal communities do not allow for spontaneous charisma, unique persona and charm; every idiosyncrasy, warts, blemishes, sweat beads and speech impediments vanish in the brilliant haze of a cartoon reality.

---

[390] Lawrence J. Terlizzese, *The Technological Simulacra*:

https://probe.org/the-technological-simulacra-on-the-edge-of-reality-and-illusion/ Jeffry Rosen, *The Naked Crowd: Reclaiming Security and Freedom in an Anxious Age* (New York: Random House, 2004); Reg Whitaker, *The End of Privacy: How Total Surveillance is Becoming a Reality* (New York: New Press, 1999); Jacques Ellul, *The Humiliation of the Word* (Grand Rapids: Eerdmans, 1985), 152.

[391] John L. Locke, *The Devoicing of Society: Why We Don't Talk To Each Other Anymore* (New York: Simeon and Schuster, 1998), 19.

The internet usurps socialization, once the privilege for home, community and school, imprinting virtue and self-esteem now belongs to the nebulous forces in media. Normalization engenders conviction in children. Modern Christians have surrendered their responsibility. Youth spend more time in front of screens and playing with inanimate devices, tickling their frontal lobes, than with adults and friends. The most technologically adapted generation is rapidly regressing into the most nihilistic since late antiquity. Media conveys materialism directly to the social body by passing critical consciousness. According to Marcuse traditional social agents cannot keep pace with modern media which only communicates, "efficiency, dream and romance . . . with this education the family can no longer compete." The father's authority is the first religious value to fall.[392]

Regrettably, educational institutions are rapidly losing relevance as homeopathic agencies where young people find partners with like interests and learn a worldview from an expert. What may be gained in convenience and accessibility for online students quickly amounts to a loss in bonding necessary for personal ownership in knowledge, discipline and character. Friendships forged in traditional settings create support groups throughout one's career. The threat to knowledge by electronic learning is subtle, but real, as an online college instructor, I did not know how to write a letter of recommendation for a student I never met in person.

People use technologies for simplicity. It is easier to send a text than write a letter. However, hasty action reveals the presence-of-absence. "I use it because it's easy," is hardly a rational defense. The trap lies here for all of us. The history of technology shows what begins with convenience accelerates into irresistible necessity. *New technology was never necessary until its invention.* What was once so easy to do because it could be done soon becomes what must be done. Fast and easy causes us to forget how to communicate in any other way. Why write a letter when you can text? Why read a novel when you can watch a movie? Why walk and talk when you can drive and listen to the radio? Why commute to class when you can sit at home in comfort? Anything worth doing requires effort and sacrifice. To raise children with an ethic of comfort, to always pursue the path of least resistance, do it the easy way, will condemn them to perpetual immaturity. Sociologists tell us that the age of maturity has grown to thirty-something for, "emerging adults." Since the industrial revolution adulthood has been prolonged from 13 or puberty in agrarian society to 18 by middle 20th century, to 24 by century's end and now age 30. The most technologically adapted are also the most ill-prepared for social life. A hurried mind dulls the spirit. Expedience hooks us into necessity by exciting the senses, rupturing the glands and

---

392 Herbert Marcuse, *Eros and Civilization: A Philosophical Inquiry to Freud* (New York: Vintage, 1955), 88.

producing stupidity and apathy that dominates our intellectual horizon. Whatever makes life easier now in the long run makes it more difficult when it fails. Speed turns into habit, habit into need and need into addiction.

An electric community is not a traditional one, in the flesh, where people can establish themselves. Telecommunication positively destroys presence. What Heidegger called *Dasein,* "being there" incarnation is absent.[393] As Comedian Woody Allan put it, "90 percent of life is showing up."[394] Ellul maintained as well, "The simple fact that I can carry a camera [cell phone] prevents me from grasping everything in an overall perception . . . it keeps me from proceeding to cultural assimilation, because these two steps can be taken only in a state of availability and lack of preoccupation with other matters—a state of 'being there [openness].'"[395] Cell phone madness along with the ubiquitous camera has become a social obsession indicative of herd behavior. "Men think in herds; it will be seen they go mad in herds, while they only recover their senses slowly, and one by one."[396] The suburban herd lives in a technological cul-de-sac, oblivious to the rest of the world. Unaware they live a programmed existence. The cellphone distracts us by creating distance between users and immediate surroundings; it closes us off to those right in front of us. This technology is anti-social. How can we effectively communicate when the most important part of who we are is no longer there? The user astroprojects to another place, she has exchanged bodily presence for spiritual vacancy. The body is there tapping away, but not the soul. The cell phone user has become a void.

Metastasis

The transfer of disease from one organ or part to another not directly connected with it.

Cellphone use exemplifies professor Baudrillard's case that simulacra has reached its final level analogous to metastatic cancer. This condition is detectable by radical transvaluation. Texting is preferred to talking however, impersonal means are least effective to the point of communication. This is true despite the apparent success during the hyperreal stage to deliver on the promise to improve the world. To make it better than what we inherited from nature and our past. Telecommunication isolates people rather than bringing them closer. Social media turns out to be intrinsically anti-social. The automobile extends our commute to work. Automation reduces jobs. The computer increases the average workload, promotes illiteracy, depersonalizes knowledge, kills privacy, and enables universal surveillance; it makes

---

393 Martin Heidegger, *Being and Time* (San Francisco: Harper, 1962).

394 Woody Allen quoted in Locke, *The Devoicing of Society,* 43.

395 Ellul, *The Humiliation of the Word*, 122.

396 Charles Mackay, *Extraordinary Popular Delusions and the Madness of Crowds* (Wordsworth: Denmark, 1995 {1841}), xvi.

pornography and violent depictions readily accessible to children. Digital dependence atrophies intelligence, logic, creative thinking and critical problem-solving skills, which causes us to forget how to reason for ourselves without an algorithm. Digitalization recreates the world in its own likeness. Instead of bestowing universal happiness, it delivers mass psychosis, exponential increase in post-industrial disorders, such as depression, anxiety, addictions, attention deficit, anorexia, bulimia, suicide and leaves us groping through an all-pervasive haze between reality and illusion. Metastasis means we have crossed the reversal point. *The law of diminishing returns* dictates that once advance reaches saturation beyond which acceleration proves unnecessary, further increase will produce opposite results than expected. The threshold was breached when modern industry, imposed technology unnecessary prior to invention. Thus, anticipated positive effects inverts and harmful consequences follow rapidly, multiplying and spreading to all social spheres — (metastasis).

No one ever asks for new technology. There was no use for much of the technology developed during the twentieth century: TV, computer, rocket, Atom bomb, tank, AK-47, innumerable gadgets galore, so use was fabricated and need artificially created. *There is no felt need for a technology that does not exist.* Technology was developed for its own sake, according to *the technological imperative;* it displaced Renaissance Humanism—*the liberation of Mankind to the glory God*—as cornerstone value and converted the golden age of science, reason, democracy and human dignity into a frightful utopia. Beneficial use becomes unquestionable and stupidly assumed because it is promoted through massive advertising campaigns and in short order another necessity is added to the burdensome litany of life's requirements. As the list for, "must have" and, "can't live without" grows, to keep pace with this lightning tempo, users unconsciously surrender freedom for self-imposed slavery, blissfully unaware of the subtle trade and untoward side-effects.

The technological condition is comparable to generational slavery. Those born in servitude accept it as normal. The, "happy slave" remains so through inability to recognize his condition. He embraces the world as he finds it with all material appetites satiated. There is no reason to protest, compounded by the fact he does not know how to; without an intellectual horizon to lift him above his reality he stays forever the same. The first step to freedom is to confess one's plight: "I am a slave!" This acknowledgement opens the possibility for change and a new direction through self-awareness. This simple act, however, proves impossible without a measure in critical reason. Conversely, technological determinism imposes inescapable destiny as natural order without a free future. We have lost our problem-solving capacities and with it the power to self-determination. We are without ability to negate the technosecular world from a metaphysical position, a higher power. We have stripped-away the only absolute reference above our limited existence and concrete situation, essential to control and subdue it to rational aims. God provides an immovable ground from which humanity can reassert itself, but the ALMIGHTY does us no good if we refuse to believe in his power or worse confuse it with the machine—the work of our hands. The awesome divine confrontation with the City of Man is convoluted through christening technological utopia, "New Atlantis."

Idolization has a dehumanizing affect, rather than amelioration, unlimited acceleration fosters more burdens and further demands; it does not provide freedom but gives us a much more complicated and dangerous world. The apostle stated simply that if we have food and shelter, we should be content.[397] Material accumulation beyond basic wants becomes another liability. Possessions own us. We have made ourselves cogs in the wheel of progress and fled human dignity with flagrant abandon. Every innovation starts with promising improvement through speed and efficiency, turning into habit, then affluent addiction. In addition to the simple requirements for food and shelter, modern life expands the catalogue to toasters, microwaves, four new steel belted radial tires for the Lexus, et cetera; freedom disappears with each artificial endowment creating an accelerated nihilism. When power emerges, unabated choice is lost ending in meaninglessness. Nihilism increases as spiritual affiliate along with technological prowess. "We know that power always destroys values and meaning," Ellul said, "where power augments indefinitely there is less and less meaning."[398] Indigence proliferates along with wealth and higher living standards reducing the scope of freedom that may differ from those ensnared in the mechanized mainstream. What possibility for a decent life remains open to those who own neither car, nor computer or possess at least a college degree? How successful will any corporate organization fare if it does not exploit telecommunication and marketing techniques, i.e. propaganda? As the world conforms itself to technical requirements it loses touch with reality outside these devices. All they know is a remade version online. The church service, for example, where believers join in bodily unity around a physical sacrament becomes an embarrassing forgery, a lone spectator in front of a one-dimensional monitor. The commandment to observe fellowship, "Not forsaking the assembly of ourselves together"[399] is impossible to keep as disembodied projections. Telecommunication prohibits assembly by abolishing the essence of community, which is incarnation. Telecasting colorful shadows for real presence inhibits the manifestation of Jesus Christ. Wherever more than one person gathers in his name he promised to be there.[400] Electric telepathy cannot substitute for spiritual reality. His apostle injured us, "if we would

---

397 First Timothy 6.

398 Jacques Ellul, *Perspectives On Our Age* (New York: Seabury, 1981), 45.

399 Hebrews 10: 25 (KJV).

400 Matthew 18: 20.

judge ourselves, we should not be judged."[401] All judgment is self-judgment. How we judge is measured in relation to how we lived according to our profession in Christ. "Be ye perfect as your heavenly father is perfect." We cast our own destiny in the judgment of others.

---

401 First Corinthians 11: 31-33 (DRA).

Intermezzo

Do you experience headaches, irritability, boredom, fatigue, loneliness and isolation without your phone or electric device, then you may need to admit you have a serious addiction? Media is the new drug and reliance is rampant. I define addiction as dependency on a substance or object we feel we cannot live without. Media acts on the brain much like a drug—a stimulus that causes a slight chemical reaction through repeated use over a period of time creates a habit, which suddenly blossoms into a full-blown addiction. The problem with addiction, electronic, chemical or otherwise is the loss of self-control and the abject servitude the addict lives according to. Self-control is the Mother of all virtues. Once we lose that cardinal quality, discipline in all areas of life breaks down, we become exclusively devoted to what amounts to an idol.

Part III

Spirit

The future of the future is in the present.

CHAPTER EIGHT The Killer Robot in the Bible

(The Image of the Beast)

Magic in one age is technology in another. Traditional magic was the precursor to modern technology both are a way to act on nature, to coerce it to do our will. Spells and enchantments were the arts and science of their day. The modern engineer plays a comparable role to the ancient magician. A robot in the bible should not surprise us bearing in mind that technology is mankind lying to himself in aspiration to godhood. Thus, Machinehead is modern counterpart to the image of the beast, the artificial construct of man. Technology, like magic, creates an illusion that fools us into thinking we're someone we're not, causing us to miss who we really are; it leads us down the path of the forbidden future. A potential world where we receive the full reward for our efforts at playing God. This technological paradise usurps traditional notions of God's earthly kingdom by replacing it with a forgery.

A Robot in the Bible?

There are more things in heaven and earth . . . than are dreamt of in your philosophy.
Shakespeare

It was common place for ancient Emperors to set up statues in their own likeness and demand worship upon penalty of death. *The Book of Wisdom* related the ultimate in distant worship when it says, "Men who lived so far away that they could not honor him in his presence copied the appearance of the distant king and made a public image of him they wished to honor, out of zeal to flatter him when absent, as though present."[402] The Hebrew children were commanded to prostrate before Nebuchadnezzar's idol or face fiery execution.[403] The Romans were no exception to this practice. No sooner was Julius Caesar assassinated then his heir apparent Marc Anthony erected an image, a statue in his honor for the crowds to gawk at, and to their awe and wonder the image could speak and move, a sorcerer's trick preformed to amaze the people. Anthony used an automaton to fool the masses into thinking Caesar had come back from the dead. However, primitive this technology was, a statue that could move and speak is upgraded to robot.

---

[402] Wisdom 14: 17 (NAB).

[403] Daniel 3.

Emperors employed this clever illusion ever since to insure their control over the ignorant rabble. Scripture alludes to this hoax when it says that the false prophet, an illusionist, "could animate the monster's image so that the machine could speak and cause all those who would not prostrate before it to the murdered."[404]

This is a killer robot in the Bible, the face of the unholy trinity in *Apocalypse* 13: the dragon, the first and second monster. The dragon is obviously the invisible power of the devil, which is the ability to craft the lie that man will live forever once he becomes a god. The first monster, literally, wild beast, is a world ruler empowered by satanic energy, a Roman Emperor in the similitude of past emperors: Antiochus IV Epiphanies, "God Manifest", the Greek Leopard, Cyrus, the Great, "King of kings and Lord of lords" the Persian Bear and King Nebuchadnezzar, "I am Babylon" the Babylonian lion.[405] The despot appeared to recover from a fatal head shot, a blow from a dagger. This phenomenal revival would cause the world to worship him—his war machine could not be defeated. His power was unprecedented in all history. He became arrogant, boastful and blasphemous. He persecuted all who did not receive his branding on their right hand or forehead: 666, the mark of a man who proclaims himself a god on earth. 7-1=6, one short of perfection repeated three times in parody of the Holy Trinity. His persecutions included baptizing opponents in pitch and setting them on fire to lighten his drunken orgies. Others were thrown to the lions for sport. Many were crucified as enemies of the state and those he liked were beheaded.

The second monster was advocate for the first; it was his job to make the first look good. We call him, "the false prophet" because he is something like a sinister John the Baptist announcing the arrival of the new man-god. As the magicians Jan'nes and Jam'bres stood against Moses by turning their staffs into serpents for Pharaoh, so he will perform miracles and great signs and wonders on behalf of the first monster. He could call down lightening to strike whoever the Emperor wished for all to see. The second monster serves the first by convincing the world the he is a god. He erects a monument to the first monster, an automaton, a statue that speaks to frighten the people and cause all who do not love it to die.[406]

In the 21st century the image of world controller looks less like a talking statue residing in the public square and more like a talking head on live TV. It appears less like the historical pattern in Nero-the-Mommy-Killer and more like a mint program adored and used by all. Orwell's Big Brother was a media composite incorporating the nations hopes and fears. This image comes alive in our collective digital veneration. We do not see one stock photo because

---

[404] Revelation 13: 15.

[405] Daniel; First and Second Maccabees.

[406] Exodus 7: 11; Second Timothy 3: 8; First Kings 18: 38.

Big Brother is a computer-generated collage. He is the projection of the grand sum of each of us constituting the global mind. No one sees it all at once because no one looks at the beast within. We refuse to take personal responsibility for our place in history. Machinehead then reflects the image of a world controller in the 21st century. Just as no one ever saw Big Brother in person or even knew if he existed, so Machinehead will project the cult of personality.

Media over exposure is everywhere. We cannot escape the city by fleeing to the country without seeing a sky-writer, hearing an airplane, or be haunted by billboards and cellphones. Media saturation produces morally ambiguous results. On one hand, advertisements provide information on how to meet basic needs, on the other, they demoralize by normalizing obscenity; not a day goes by without witnessing mass sacrilege, objectification of women, unconscionable exploitation of addiction, valorizing gambling, drug use and alcohol. Certainly, we are smart enough to feed, house and cloth the world without appealing to the lowest common absurdity.

The modern blueprint for slavery and totalitarianism is to meet all material needs, while simultaneously limiting intelligence among the proletariat, making commoners just smart enough to take orders and get the job done. Slavery is a structure of consciousness and consciousness determines who we are. Media programing restricts mental acuity leaving users without the ability to think for themselves. Mass media offers no possibility for dialogue and social media is a parody of democracy. They recreate history and reality to fit a consumerist's mentality. All this communication technology goes into *amusement*, literally, to cause not to think; ignorance is bliss. According to commercials, nirvana is achieved by possessing one of every consumer object on the market, and the forbidden fruit was really a candy-bar, it's fun to get drunk and lose control and cartoons, the children's medium, pushes pharmaceuticals.

Mass media in Oceania consisted of little more than popular entertainment (to hold attention agreeably with amusement), its litany included: racy tabloids, dime novels, film, radio, and all pervasive TV screens, with hidden cameras everywhere, whose content was strictly comprised of campy material covering a limited range of topics, not excluding a grotesque preoccupation with sex, crime, violence and astrology, we might say, "meteorology." Sports, of course, the ultimate distraction took center stage. Music was sentimentalism, electric swoon, produced by machines, not unlike the contemporary techno sound made without instruments, the imitation of music. In *1984* language was castrated by word reduction. Political control was reinforced by editing vocabulary and replacing traditional words with artificial ones. Most of these alternates eliminated complementary opposites in thinking; for example, the dialectic between light and dark was recast by striking the word, "light" and substituting the inane, "undark."

Word reduction and the predominance in picture learning are two allies that have a stultifying impact on public perception making people scarcely aware they are being manipulated by a totalitarian regime. In the case of our technosecular environment subtle control proceeds not from tangible public representatives, rather we are seduced by our own desire for speed, order, ease and convenience. We have summoned the technocratic order. We are haplessly addicted to the premise that faster is easier, easier is better. We hardly realize that we have sold our souls for a better mousetrap and a richer cup of coffee. Like Esau trading the family birthright for a hot bowl of lentil soup, we barely notice what we have lost in exchange for instant gratification.[407]

Visual media ends free thought by telling us what to think. The image makes a final statement by whatever is pictured. Literary imagination never regains poise after watching a movie. When we see *Star Trek* or *Star Wars*, we cannot think of the future in any other light than fantastic speed. Mass media is the ultimate mind control machine. Reduce words. Abbreviate thought. Curb freedom. Show the people what to think. Program their reality according to pleasurable thought or vamp things up a bit when they become stagnate and in need of stimulation. Philosopher Karl Jaspers makes this frightening comment on the falsity of a visually framed reality, "Pictures deceive us by their visibility and vividness, and reality itself is changing fast: a comparison of travel accounts from the past century shows that what was once real yesterday is no longer real today."[408] Images are a falsification of reality. They distort reality, then, claim to be depicting an accurate representation of it.

Yet, the be-all and end-all in totalitarianism is the barbarization in public morality created by the death of privacy. Orwell was more accurate than we have the courage to take seriously,

> With the development of television and the technical advance which made it possible to receive and transmit, simultaneously on the same instrument, private life came to an end. Every citizen or at least every citizen worth watching could be kept for twenty-four hours a day under the eyes of the police and in the sound of official propaganda, with all other channels of communication closed. The possibility of enforcing not only complete obedience to the will of

---

[407] Genesis 25: 29-34.

[408] Karl Jaspers, *The Future of Mankind* trans., by E. B. Ashton (Chicago: The University of Chicago Press, 1961), 96.

the state, but complete uniformity of opinion on all subjects, now existed for the first time.[409]

A world without privacy is a world without shame. A world without shame possesses no virtue. Society without virtue is democracy run amok. Such cultural chaos is the incubator for tyranny.

The Grand Illusion

Don't be fooled by the radio
The TV or the magazines
They show you photographs of how your life should be
But they're just someone else's fantasy . . .
America spells competition, join us in our blind ambition
Get yourself a brand-new motor car
Someday soon we'll stop to ponder what on earth's this spell we're under
We made the grade and still we wonder who the hell we are.[410]

Sinful-Humanity will bring prophecy to climax by embodying rebellion. We play-God by assuming the powers of God; the Apostle declared concerning the World Controller that he will exalt himself, "proclaiming himself to be God."[411] This supposes the perennial temptation to usurp God's place and reveals the secret to lawlessness. Listen to Reformer John Calvin explain the deceitful origins to our temptation to godhood. "[Mankind] cannot arrogate to himself one particle [gene, atom, molecule, *etc.*] beyond his due without losing himself . . . whenever our minds are seized with a longing to possess a somewhat of our own, which may reside in us rather than in God we may rest assured that the thought is suggested by no other counselor than he who enticed our first parents to aspire to be gods."[412] This is the ultimate *coup d'état*; it prefigures the coming of, "The tyrant whose presence is infused with Satanic energy working through the forsaken by powerful miracles, strange phenomena and much slander," the Apostle briefed us, "because they did not receive the truth in love to be saved," he warned, "so then God will turn them over to the grand illusion believing the lie, so that, they may all be judged who do not believe the truth but revel in wickedness."[413]

The grand illusion is the lie that we can effectively play-God without losing ourselves. I simply advert playing-God is not worth the risk. We are delusional to believe technology frees us to be masters of the universe, closer to truth we will destroy the earth in the liberation process. The cyborg condition will be exchanged for the human one. We debase ourselves in

---

[409] George Orwell, 1984 (New York: Signet, 1949), 169-170.

[410] Styx, "The Grand Illusion" *The Grand Illusion* (A&M, 1977).

[411] Second Thessalonians 2: 4 (RSV).

[412] John Calvin, *Institutes* II, 2 ,10.

[413] Second Thessalonians 2: 9-11.

deference to a superior race. Computer Science professor Matthew Dickerson prophetically asked, what if the transhuman, "transformation is based on something that is not true?"[414] Into whose image are we transforming? It is an equally formidable task to decipher the image of man for man as it is to decode the meaning of the image of God in man. This problem is compounded when we try to discern that image in light of the atom bomb and AI. "The heart is deceitful above all things, and desperately wicked: who can know it?" the prophet declared.[415] Theologian Reinhold Niebuhr said, "It is not unfair to affirm that modern culture, that is, our culture since the Renaissance, is to be credited with the greatest advances in the understanding of nature and with the greatest confusion in the understanding of man."[416] The will of Man is more arduous to understand than the will of God. To the question of what end Expressionist Vincent Van Gogh painted he answered, "je ne sais quoi" (I know not what).[417] This is the mystery of faith the secularist does not fathom. Medievalist C. S. Lewis instructed that when humanity prevails over nature's fury we subject the entire race to the caprice of a handful of, "individual men" and, "those individuals," he said, will lose themselves to what, "is purely 'natural'—to their irrational impulses."[418] Frank Herbert in his famous novel *Dune* echoed Lewis, "Once men turned their thinking over to machines in the hope that this would set them free. But that only permitted other men with machines to enslave them."[419] Men with machines are not liberators, but slavers. The device, agent that frees from limits, is the vice that holds us in place. The computer screen mediates the world reframing reality making it smaller. So, the cyborg will diminish human life by making it obsolete and inferior to itself.

Freedom encompasses three forms: religion, reason and community. Humanity is religious by nature. As creatures we identify with a purpose greater than ourselves. Faith is a distinguishing mark of humanity. Intelligence affords us the ability to make choices according to a higher principle, which gives meaning and coherence. This is freedom grounded in Being. Aristotle said, "man is a social animal." People have an innate connection with a larger group that provides a sense of belonging, stability and purpose. We are communal: husband, wife, child, tribe, language, and nation; just as God is a

---

[414] Matthew Dickerson, *The Mind and the Machine: What it Means to be Human and Why it Matters* (Grand Rapids, MI: Brazos, 2011), xiv.

[415] Jeremiah 17: 9 (KJV).

[416] Reinhold Niebuhr, *The Nature and Destiny of Man,* V. I. (New York: Charles Scribner's Sons, 1941), 5.

[417] Vincent Van Gogh quoted in Gabriel Vahanian, *The Death of God: The Culture of Our Post-Christian Era* (New York: George Braziller, 1960), 182.

[418] C.S. Lewis, *The Abolition of Man* (New York: Macmillan, 1947). 79, 80.

[419] Frank Herbert, *Dune* (New York: Ace, 1965), 11.

family of persons: Father, Son and Spirit. People remain in fellowship with others ordering life according to his higher power and care for creation.

By establishing human *Being*, we prove the existence of God. Our own existence testifies to the Creator. We reflect the divine image, even if only partly. We are a synthesis between infinite and finite. "Infinity in finite form."[420] Any attack on the human shape is a reproach to the character of God. The denial of man involves the denial of God. We are only fully human in God. God is relevant to the world only when we believe in his existence. A dead god is a mockery and means the idea of God has lost relevance in a world that is self-sufficient. God is dead to a world that no longer experiences his presence. The existence of God outside the divine-human encounter, regardless of its monotheism is idolatry. God exists in relationship with us, or he doesn't exist at all. Only in God do we transcend nothingness as Lord of Earth. Without God to give purpose to the world all values flatten because there no longer remains anything greater than ourselves. We are free to be ourselves only if God exists. Idolatry requires submission to an order lower than ourselves. Faith raises us to our true selves beyond the world but located directly in it. Without the counterbalance in threefold division in human nature we move freely in strength without wisdom and without metaphysical and ethical limits which is chaos.

Technology presupposes life. A higher purpose for which it was designed; however, it fosters perverse nihilism by usurping that sacred end with its own cause. Technology for its own sake is knowledge that kills the user. Research and development without reverence for life produces the hydrogen bomb. The final limit to all horizons. As a child during the Cold War I identified with Architect Richard Ross' feelings concerning what life was like for youth living in the shadow of the nuclear arms race, "by the time I was older enough to start planning a future, I was pretty sure I wouldn't have one. I am still amazed I got to grow up. Amazed!"[421] Albert Einstein said, "The atomic bomb has changed everything except the nature of man."[422] The invention of the atom bomb and the absolute necessity for its abolition is the defining element for all time. If we want a rallying point for a prosperous future, if we care about the lives of our children's children, the Earth and the world to come, we cannot entertain human extinction. Philosopher Karl Jaspers stressed the importance in finding a solution to the global problem raised by unleashing the nuclear genie. "Without world peace," he said, "there

---

420 Nikolai Berdyaev, *The Realm of Spirit and the Realm of Caesar* (New York: Harper, 1952), 36.

421 Richard Ross, *Waiting for the End* (Princeton, NJ: Architectural Press, 2004), 12.

422 Albert Einstein quoted in Paul Boyer, *By the Bomb's Early Light* (New York: Pantheon, 1985), 36.

is no preventing the extinction of mankind in an atomic holocaust."[423] Why work toward a future void of people? How can we live comfortable today knowing we will execute our children tomorrow? Why fight the future when we can change it? Everything we do has significance, no matter how small or great. With each joyful smile we inch away from the abyss. For good or bad we hold the power over the future in every decision regardless of how private, trivial or politically grandiose. "Now, unless all of us live with and for one another, we shall all be destroyed together," Jaspers remarked soberly. We do, "in miniature, what on a larger scale makes mankind destroy itself."[424] Whatever ethic causes people to rise to the occasion in times of crisis must be woven into our daily fabric to be effective. People face trials and tribulation the same way they do the normal conditions of life with courage and fortitude or in fear and ignorance, as in life so in death. The truth remains one and always the same, yesterday, today and tomorrow.

Limitless behavior ends in self-destruction. Extinction looms over an indefinite technological horizon like the Sword of Damocles, threatening any attempt to refit ourselves for immortality with a grand explosion, painful asphyxiation and subtle replacement. Albert Einstein said, "radioactive poisoning of the atmosphere and hence annihilation of any life on earth has been brought within the range of technical possibilities."[425] Astrophysicist Stephan Hawking recently proclaimed AI and genetic engineering menaces to global survival over the next 100 years, "I think the development of full artificial intelligence could spell the end of the human race," he said frankly.[426] *SpaceX* founder Elon Musk exclaimed concerning AI, "We are summoning the Demon."[427] And partner in *DeepMind* Shane Legg said bluntly, "I think human extinction will probably occur, and technology will likely play a part in this."[428] A limitless future is never possible for a limited creature regardless of how advanced. The biological needs to eat, sleep and procreate are guidelines placed on us to curb excess. Death is the final limit, without which new life and growth will not occur. A thousand-year Reich was possible if the regime's fathers never died, effectively murdering all further progress. Tomorrowland is a lie for those who believe that death is the final obstacle to immortality

---

423 Karl Jaspers, *The Future of Mankind* trans., by E. B. Ashton (Chicago: The University of Chicago Press, 1961), 16.

424 Ibid., 24, 25.

425 Albert Einstein quoted in Jonathan Schell, *The Fate of the Earth* (New York: Avon, 1982), 12, 13.

426 Stephan Hawking quoted in, "Rise of the Machine" in *The Dallas Moring News* (Sunday, February 14, 2016), 1P.

427 Elon Musk quoted in, "Elon Musk's Future Shock" *Vanity Fair* (April 2017), 118.

428 Shane Legg quoted in, "Elon Musk's Future Shock," 118.

resolved ultimately upon resurrection.[429] To believe that death can be overcome with enhancement will create an abomination.

Technology that destroys life kills itself. Regression began when progress lost its spiritual ground with the advent of history's first technological singularity at the dawn of the nuclear age. For the first-time humanity acquired the means of its own annihilation. We effectively acquiesced to extinction. Closing any further intellectual horizons. The imminent convergence of ICBM's and AI is flirting with disaster. No further progress is possible until we resolve the approaching danger unleashed by the Manhattan Project, the near simultaneous invention of the computer and rocket technology. Fascination with Mars is a distraction that fails to account for the hypothesis that the reason we do not encounter extraterrestrial intelligence, is that, it is not possible for civilization to advance beyond its industrial phase without destroying itself. Faster is the future, yet we remain the same. Since 1945 acceleration has been bound to a perpetual repetition in the present order. The future is no different today than yesterday, only faster. Efficiency as guiding light has not abated since the computer's first Parousia, but has accelerated the future's arrival as fixed destiny, while nothing essential has changed. Speed fastened in place the cardinal value. Accelerated time has no purpose except for the hell of it, an irresponsible attitude that beckons the day of reckoning. No universal consensus prevails on the rush to the future, not even human improvement. Mankind as presently constituted is not on the agenda; otherwise we would not be talking about a posthuman singularity. I hear no universal definition for "better" other than "faster." When liberation ceases to be the goal for progress we have crossed the Rubicon. The law of reverse effects dominates and transvaluation pervades all. We now serve the computer not *vice versa*. Liberation will bring free choices according to self-determination creating a meaningful universe because it will be one in which we feel responsible, where we can affect the course of change through our higher power. The fact that we have a choice between 31 flavors of ice cream and from red, white and blue cars, but we cannot disarm a hoard of nuclear warheads, is not liberty, but the illusion of freedom on parade. A hypnotic spell cast by the media grid further entangling our loyalties with instant gratification not spiritual satisfaction for a life well-lived grounded in the mystery of faith. Faith without a center is anarchism leading to nihilism; it is zeal without knowledge going nowhere fast. The illusion of progress spawned by mass media hides the reality of our regression into impending oblivion.

The posthuman refusal to define human nature in male and female distinction above the world does not open our horizons. We cannot create ourselves without presupposing a greater definite center. We are left

---

[429] First Corinthians 15.

defenseless and exposed regressing into a strict rarefied scientific order as final reference devoid of God. Media closes the future through eternal recurrence. The same materialist theme: buy, drive, fight is repeated endlessly. *Space the final frontier* is a fiction that prohibits alternative conceptions. Contemporary thought on the future is limited by the fast-foreword version predominant today; more of the same just faster. This means the twilight of mankind, the end of history, frozen in whatever petrified lump, the second technological Parousia, i.e. singularity, will find us. Our final form will be indistinguishable from our Paleolithic ancestry up from Africa, a brutish imp.

Simulacra replaces reality with a digital copy made in our own image—the copy of a copy. The duplicate, of course, is never greater than the original. "The disciple is not above his master, nor the servant above his lord," Jesus said.[430] Man represents God, the earthly prototype to the heavenly Father, the *imago dei* is the glory of God on earth; art captures him in a technological medium, which is religion, it is that facsimile that provides the outlook for how we view ourselves through the digital prism— the copy of a copy. To believe that a glorified reproduction, a snapshot somehow surpasses the unique person demonstrates how far along the popular illusion has accrued. Media qualifies as a delusional influence by remaking us in the likeness of a simulation creating a soulless drone that is blind, deaf, dumb and sterile. Media technology's bombastic proclivities silences its users. The drone equals a deconstructed humanity imitating God. A divine resemblance once removed that recreates homo sapiens as appendage in abbreviated form. Believing in the Machine-god, literally repeats the biblical model for idolatry. The Transhumanist enterprise fashions an ideal, the cyborg in the form of a replicant exalting itself to godhood, but brutal subjection places us lower than our own design. Man builds a machine, he feels represents God, which in truth is a replacement for himself in the likeness of a clone less than God. He lowers himself to a subhuman thing that appears real but does not exist outside faith in his own highly-stylized self-projection.

The image of God means God comes to earth in the birth of a child; however, we degrade that likeness by projecting our highest qualities along with our greatest vices onto a machine. The semblance between God in heaven and man on earth diminishes to a fearsome shadow lower than both. The Artilect will always be inferior to us, but we are entrapped the instant we believe the creation greater than its Maker.

---

430 Matthew 10: 24 (KJV).

The Forbidden Future

I've seen the future and I've left it behind[431]

Technology that pushes the frontiers at risk to humanity compels us to the forbidden future; a potential world where unwarranted curiosity seduces us into an irretrievable chain of events that would have otherwise been avoided. Extinction, in the words of writer Jonathan Schell, is "a human future that can never become a human present."[432] Any endeavor that sacrifices survival for instant gratification is obviously absurd. Immediate gains that endanger posterity are self-defeating and therefore irrational. When there is more to lose than there is to gain wisdom counsels a different path. It was believed during the nineteenth century that advances would raise society's spiritual condition automatically, so that there could never be any discrepancy between morality and progress. Science would solve ethical quandaries by improving the material world: feed the poor, cure the sick, house the needy and abolish war by beating plowshares into swords. The fundamental contradiction in modernism is that technological progress will inevitably lead to peace not war. The Maxim Machine Gun and the British Dreadnought were super weapons believed to make an army or navy invincible, unassailable making challenge unthinkable—ending warfare. This logic is repeated today, we know, however, that efficiency in killing does not make war less likely. We cannot end war by building greater and more horrible weapons. Proliferation leads only to genocide.

Technology is a means to an end, but when the end justifies the means, the agents for progress exceed their limits. This is the will-to-power, as Berdyaev declared, "The will to power is the will to murder . . . the will to power can only be realized through murder."[433] Science is murder when grounded in power and renounces reverence for life. It loses moral authority as humanism's vanguard, turning into its radical opposite, posing the greatest threat to survival by treating people like gambling chips thrown down in pursuit of convenience. This was the case with the atom bomb. The Trinity test-explosion in 1945 was the early design towards genocidal technology. To venture life for the unknown, to risk the world for knowledge, to chase forbidden fruit is to acquire dark-light under the guise of enlightenment contributing to our fall from grace. To prevent injustice instrument and purpose must cohere. Like bad company corrupting good morals, so unjust channels produce unholy causes by giving rise to anarchy, which leads to

---

[431] Black Sabbath, "Supernaut" *Black Sabbath Vol. 4* (Record Plant, 1972).

[432] Jonathan Schell, *The Fate of the Earth* (New York: Avon, 1982), 138.

[433] Nicolas Berdyaev, *Towards a New Epoch* (London, UK: Bles, 1949), 9; Second Thessalonians 2: 7

tyranny, law's end. Tyranny is anarchy come to power. And since the atom bomb the problem of power has eclipsed all others.

Knowledge is power encapsulates the technosecular belief system. The prevailing nihilism was not supposed to happen; although it was entirely predictable. *The law of regression asserts that as power increases choice decreases.* A global techno-order invariably destroys any independent spiritual existence. Voter apathy prevalent today is a good example. The ballot appears useless in effecting change because volition is coerced by limiting the range of candidates and issues. "Free . . . and meaningful choices are eliminated. They simply cannot co-exist with the necessity of efficient planning [campaigns, TV ads]."[434] The Faust legend teaches that power comes at the price of your soul. Jesus was emphatic about the obstacle materialism presented to salvation. "What good is it for someone to gain the whole world, yet forfeit their soul."[435] Likewise, the Buddha announced, "It is the law of humanity that though one accumulates hundreds of thousands of worldly goods one still succumbs to the spell of death."[436] Power's depraved influence does not cease when it changes accent from politics and religion to science and technology corrupting its handlers after eradicating dissent, so technological over-extension turns users into victims. The conquest of nature leaves the eclipse of God and man in its wake. Total systems control creates a void, the abeyance of negation. Necessary boundaries that provide spiritual compass disappear making restraint impossible. Negation is the path to limits and revival. A world that knows no fear proceeds with great know-how, but no wisdom. Engineers rush in were poets fear to tread. Nietzsche understood the importance for limits. "Wisdom sets bounds even to knowledge" he said. And T.S. Eliot confirmed, "where is the knowledge lost in information? Where is the wisdom that is lost in knowledge?"[437]

The technological future is predictable because it races toward a predetermined end established by its own inner logic. Like Oedipus Rex we cannot change our destiny through cunning reason, which ensnares us further. In trying to avoid fate we seal it by giving it credence. Knowing future events insures their arrival. Like Ebenezer Scrooge we ask the Ghost of Christmas Future upon witnessing our own tragic demise are these visions of what, "Will be" or, "are they shadows of the things that May be only?"[438] The future is worse than useless if we cannot envision more than one potential outcome.

---

[434] Lawrence J. Terlizzese, *Trajectory of the 21st Century: Essays on Theology and Technology* (Eugene, OR: Resource, 2009), 47.

[435] Mark 8: 36 (NIV).

[436] Buddha quoted in *Jesus and Buddha: The Parallel Sayings*, Ed., Marcus Borg (Berkeley, CA: Ulysses, 1997), 71.

[437] Nietzsche and Eliot quoted in Terlizzese, *Trajectory of the 21st Century,* 131.

[438] *A Christmas Carol: by Charles Dickens; illustrated by Roberto Innocenti* (Italy, 1990[1843]), 135.

If we disallow for an irrevocable course, we must allow for a conditional one, open to difference and subject to change. Inevitable returns exist when we decline to vary cause and effect. We stubbornly maintain the present trajectory. We change the future by changing ourselves. Scrooge remade himself for selflessness, to care for the needy, feed the poor, to love the unlovable. Similarly, Admiral Kirk refused to relent to the confines of the possible, when as a cadet on the authority of faith he changed the *Kobayshi Maru* simulator by reprogramming the computer not to permit a lose-lose scenario. "I don't believe in a no-win situation," was his only defense. Ambitious faith is needed to counter our hopeless belief in materialism. If logic dictates that certain happenings are inescapable, a no-win dilemma, a lonely death, patricide and incest or doomsday, then, these futures can be altered through *causal reversal.* This removes the causal agent in the cause and effect sequence. The absolute inevitable becomes the conditional. Things that were once *must be*, convert to *may be.* This is a simple, although not an easy process. We are responsible for self-control all other power is illusory. We can spend ourselves on stupid fleshly pleasures or commit to saving the future by reversing our direction today.

Technology was once liberator, freeing us from stifling tradition and unremitting nature, old world order; it is now characterized by the opposite. The doctrine on limited government that defines the modern Republic from old world absolutism cannot preclude a new ethic surrounding technological limits. Government over-reach and technological acceleration are co-extensive and interdependent. Unlimited returns will be the end to democracy as the recent death to privacy proves. In the nineteenth century technological nerve replaced liberation as core belief. The Industrial Revolution and Manifest Destiny overpowered Jefferson's simple agrarian republic. So long as technology remains the bases for progress, "the rest will care for itself," the automatic hand of techno-providence will direct our culture. Instrumental value remains the all controlling motif; freedom, equality, harmony, beauty and the pursuit of happiness revert to secondary status—nominal at best.

Still the Ghost pointed downward to the grave by which it stood.
"Men's courses will foreshadow certain ends, to which if persevered in,
they must lead," said Scrooge. "But if the courses be departed from,
the ends will change. Say it is thus with what you show me!"
The Spirit was immovable as ever.[439]

[439] Ibid., 135, 138.

Techno-Millennialism

Any sufficiently advanced technology is indistinguishable from magic.[440]

A machine is a slave, a thing that has no right to exist free from whatever alien purpose its user may wish. The classical world asserted a strong aversion to mechanization because it was believed to be centered in slavery. Mechanical tools belonged to craftsmen who were mostly slaves. Athenian humanism associated technology therefore, with slavery as demonstrated in the Periclean Age, which excelled in all manner of arts and science, except technology, which remained slave work. "The slave was quite literally, man reduced to machine."[441] For Plato mechanical learning required no creative thought since its operations are consistently automatic. Aristotle's doctrine on causes remains relevant, he argued for four causes that bring ordinary things into existence, a chair, for example, has a *material* cause, which is the unfinished wood that serves as raw material. Second, the *formal* cause would be the idea of a chair; then, there is the *efficient* cause, the tool used to construct the chair, a carpenter and his hammer are not as we might suspect, essentially different. Lastly, the *final* cause is the actual chair. Despite the plethora of imagery used in literary analogy, the ancients considered technology to be imitation of nature existing on an inferior level to the ideal concept. The organic stood above the inorganic. The chair was not greater than the tree it was hewed from. The statue was below the human form it represented. *Man is the measure of all things.*

The machine order does not receive religious sanction until the middle ages, when it becomes the means for redemption. Historian David Noble pointed out that from the French Emperor Charlemagne to early modern times technology was associated with salvation. It was believed Earth could be returned to Edenic paradise through technological progress and the lost image of God would be restored.[442] Theologian Ernst Benz, like-wise taught that the modern project was founded on a theological notion in which man believed himself to be a fellow worker with God in establishing the kingdom of heaven. Technology reverses Adam's fall.[443] Millennial theology

---

440 Arthur C. Clarke quoted in *Profiles of the Future: An Enquiry into the Limits of the Possible* (n.p., 1973), 36.

441 Aram Vartanian in "Man-Machine from the Greeks to the Computer" in Philip P. Wiener, ed., *The Dictionary of the History of Ideas*, vol., III (New York: Charles Scribner's Sons, 1973), 134.

442 David Noble, *The Religion of Technology* (New York: Knopf, 1997), 9.

443 Ernst Benz, *Evolution and Christian Hope: Man's Concept of the Future from Early Fathers to Teilhard de Chardin*, trans., by Heinz G. Frank (New York: Doubleday, 1966), 124-25; Lawrence J. Terlizzese, *Into the Void: The Coming Transhuman Transformation* (Cambridge, OH, 2016), 20, 21.

overemphasized the human role in creating the New Jerusalem turning into a works-based salvation.

In the Patristic era, the so called "mechanical arts" were never conceived as a means to Edenic restoration; rather theology shared the Platonic view that held technology to be intrinsically bound to the material world. They could have no redemptive qualities. Super secular science today belies its roots in Millennialism. Noble summarized, "For modern technology and modern faith are neither complements nor opposites, nor do they represent succeeding stages of human development. They are merged and always have been, the technological enterprise being at the same time, an essentially religious endeavor."[444] Technological progress was expected to reverse the effects of the Fall and recreate original perfection. Transhumanism extends this vision into the twenty first century. However, a thousand years into this Millennial venture, so to speak, we do not find a regenerated planet ruled by the Son of David as the prophet foretold.[445] Instead, we encounter a world controlled by urban powers assuming divinity. We find a society that grows at the expense of irreplaceable natural resources, like a huge war engine burning all it can as fast as possible (Nahum 3).

A slave is a machine, legally and effectively. The ancients would be scandalized at the cyborg condition. They might see it as something like the revenge of the slaves. The human-machine hybrid was not a viable concept until the eighteenth century, when it received considerable traction in connection to the Industrial Revolution. It serves as an intellectual framework for a mechanistic future. Tech goes from subhuman slavery to superhuman mastery and from less than nature, to more than man. Technology, however, is neither animal nor God, but completely man-made, a contrivance. Whatever we think human nature to be, it is that concept that gives rise to our machines. Technology is never any more or less than who we are.

---

[444] Noble, *The Religion of Technology,* 4,5.

[445] Isaiah 11.

CHAPTER NINE Millennials and the Singularity

Old ideas die-hard and sometimes they come back from the dead. The Law of Inevitable Progress is one such idea. From one light among many, it has become keystone idea in Western civilization, and from ridicule among a plethora of posties by the end of the 20th century, it has experienced reanimation in our day. Technological acceleration is the one absolute around which all else centers. This optimism fails to grasp the negative ramifications, so called, "progress" may have and that regress is just as characteristic of our experience. This is especially true in terms of warfare. The fact that technological improvement makes the means of war more-deadly is cause for us to question the trajectory of the entire project. Knowledge is the basis for technological advance, but when that knowledge proves opposite to the interests of life the whole enterprise becomes suspect.

Absolute Technology

Can human beings make their own gods? These are not gods at all!
Jeremiah 16: 20
(NJB)

Technosecular society is by no means an amoral relativist age as was often suggested by Postmodernism. *Post* means *after*. This says absolutely nothing about the society it wishes to describe. A Postindustrial age reveals only that we do not know how to describe the times we live in, just as Posthumanism reveals our ignorance about what life form is emerging from our technology. Techno values are solidly fixed and implacable. Traditional values recede in the face of the technical juggernaut, like a bulldozer it levels all in its path eradicating the past. It has no need for a new synthesis. Our techno age distinguishes itself from every other one by the predominance in its greatest value, the technological imperative: what can be done must be done. This value is demonstrable through technological acceleration. This is not, "progress," which assumes improvement in the human condition. Historian Arnold Pacey informed us that, "Progress in deed is a misleading word to use about technology, because it implies steady advance toward a single, universally accepted goal. In fact," he said, "the direction of technical progress has varied in the past. The choice of direction is always open . . . it is important to appreciate the full range of options that are available."[446]

Techno order first appeared in Western civilization and then through modernization, its values and infrastructure were transmitted to non-Western countries. The world grows smaller and more uniform through the elimination of space and time. Technology replaces natural limits and cycles. There was no need for the great innovations since the nineteenth century, such as electricity, phones and cars. The creation of new technology generates its own need. All skepticism is ruled out. Criticisms are never disproven or reasoned about, they are simply dismissed as nonsense, "of course," is the standard response, "society is better off with electricity, TV and vaccination." Propaganda will even suggest that the sanity of the naysayer may be in jeopardy. Theology affords metaphysical certainty to new technology arguing that, "electricity is a gift from God, a blessing." Theology supports technology in the same way as it does the State with the doctrine of the divine rights of kings, giving innovation an unimpeachable character. Technology replaces the past by overturning all traditional order. As the next generation is born into the new ways, they are accepted uncritically. There will no longer be any living memory of a time without electricity, phones or cars. Technology is absolute and inescapable.

Decline Verses Acceleration

Write it on a tablet they can keep, inscribe it in a record:
That it may be in future days an eternal witness.
Isaiah 30:8
(NAB)

Social philosophers have long hailed the demise of Western civilization since the nineteenth century; this begun with the Malthusian grim prediction surrounding the exhaustion of natural resources in the face of overpopulation. Then continuing with Jacob Burckhardt, Friedrich Nietzsche and Oswald Spengler, all felt the drain from a mechanistic worldview and ensuing loss of vitality.[447] Ennui had set in these cultural critics recognized a consistent outline in the decline of society. Three categories remain the same throughout this pattern. First, *the loss of traditional values*; values that once gave society its cohesion, such as patriotism, family and religion begin to recede and fall by the wayside. Second, *the decline in native population* and *influx of aliens* predominate in politics. Demographics shift in favor of immigrant populace with a sharp decline in the indigenous peoples. And thirdly, *decadence in art and technology* becomes evident in military ventures and public work

---

[446] Arnold Pacey, *The Maze of Ingenuity: Ideas and Idealism in the Development of Technology*, 2nd Ed., (Cambridge, MA: MIT Press, 1992), 27.

[447] Arthur Herman, *The Idea of Decline in Western History* (New York: Free Press, 1997).

projects. The traditional signs for decay hold fast for the 21st century, except for advance in technological acceleration and military exploits. Technology is the only element in modern culture that continues to climb while the normal cultural centers such as family, religion, art and civics recede. Jaspers confirmed this point, "Today," he said, "the most conspicuous difference between our own time and the third century A.D. is that then technique was stationary or retro grade, whereas now it is advancing with giant strides."[448] The approaching singularity is all about technology.

The cyborg condition is not an improvement on the human one, but its abolition. Man's faith and love are waning comparable to rapid acceleration leading to singularity, a time when our consciousness will be indistinguishable from artificial intelligence and history will culminate into a planetary being. What biophysicist Gregory Stock called, "Metaman" a superorganism, in which individuals exist as solitary cells in the larger social body, a Leviathan, an artificial collective man according to philosopher Thomas Hobbes.[449] The only ones who stand to benefit from this, "rapture of the nerds" as it has been sardonically called, are those wired into the system, who will appear to evolve to the next level, which is yet unknown.

One possibility is an evacuation from the body that translates the mind into a digital existence. Something like what we might see in movies, such as *Star Trek: The Motion Picture*, 1979, or *Lawn Mower Man*, 1992 and *Transcendence*, 2014. The digital rapture paradigm is modeled off the Gnostic Ascension motif: the soul transmigrates into higher realms without the encumbrance of the flesh. Initiates will be free to roam the celestial spheres inside the confines of a computer server. I believe this will be a step down. What kind of progress can be made without the body? Posthumanism will prove to be a monstrous dehumanization. At the very best one will emulate attributes of consciousness, speech and brain patterns, voice recognition and the like. Consciousness, uniqueness, or soul cannot be cloned. Another possibility is that a radical change in the body will occur through the addition of an exoskeleton with genetic modification not unlike the film *Elysium*, 2013, which depicted remaking the body or pseudo-resurrection as a pale transfiguration. An irrevocable fusion will take place creating an enhanced version. The digital will join with the body announcing the dawn of the cyborg age. These are permanent states of being. They can get no higher. Just as contemporary humanity sees primates, such as chimps and apes, so future Avatars will consider those left behind as cute taking monkeys. The singularity represents a parodical belief in progress, where acceleration leads to an absolute zenith, a critical mass that explodes, "the full range of options"

---

[448] Karl Jaspers, *Man in the Modern Age* (New York: Anchor, 1951), 21.

[449] Gregory Stock, *Metaman: The Merging of Humans and Machines into a Global Superorganism* (New York: Simon and Schuster, 1993).

progress makes available. The singularity limits our purposeful direction to one end: the cyborg. What will become of the rest of humanity that remains unplugged?

This is not progress, but regress because it forgets the cosmic order: *the greater does not come from the lesser.* The student is not above his teacher mankind cannot make a being greater than himself any more than God can make a rock too heavy to lift. Technology assumes life and the extension of mankind not its opposite. "Every meaningful action presupposes the continuance of life," Jaspers declared, "to contradict this thesis seems antihuman."[450] Inhuman because, it says, Posthumanism has lost faith in mankind's irrepressible drive for a meaningful life. Mankind is no longer worthy to live as anything but a slave, an engineer, to tend the machine. Human potential is limitless, but we will never reach it if we curb our own prospects. "Man, either grows in freedom and maintains the tension of this growth or he forfeits his right to live," Jaspers said, "If he is not worthy of his life, he will destroy himself."[451]

The Artilect War

War is Hell.

Singularitarianism believes that the advent of a super intelligent being will ultimately prove beneficial, but fails to weigh the potential negative consequence such a god-like creature will cost us. Critics pose a different fate for mankind. The Artilect War looms over the 21st century like a dark cloud threatening to rain death upon the world's population centers. AI researcher Hugo de Garis predicted that by the end of the century a major war will erupt between Cosmists and Terrans over the creation of Artilects, artificial intellects. This war will involve the death of billions measured as gigadeath. If one extrapolates the body count in modern wars from the Napoleonic Wars about 5 million through the Great War 15 million to World War Two 66 million, we eventually come to death in the billions by the next war. This war will be fought with nuclear weapons and other advanced 21st century technology.

The war will be waged over the construction of Artilects, which are massively intelligent computers which will exceed human intelligence trillions of times over according to de Garis. These Artilects will threaten human existence because they will perceive us as a nuisance much the way people consider insects pests and wish to exterminate them. Their construction will be possible towards the end of this century. A species dominance debate will

---

[450] Jaspers, *The Future of Mankind,* 166.

[451] Ibid., 183.

direct the course of the 21st century, which asks, "Do we build gods, or do we build our potential exterminators?"[452]

Over the next two decades computers and robotics will increase in intelligence and proliferate everywhere. Artificial intelligence will first appear primitive and stupid compared to humans but will continue to grow exponentially until it rivals our capacity. Machines will become teachers, babysitters, friends and household servants. There will be teacherbots, homebots, even sexbots. Most people will become alarmed at the increasing ability of AI and will want to limit its growth before it exceeds humanity's. People will divide into three camps: Cosmists, Terrans and Cyborgs. The first group takes a cosmic view of AI turning the construction of Artilects into a new religion. They worship the machine. They believe it represents the next evolutionary stage. Humanity becomes God-Makers. Cosmists will risk human extinction as the price for their construction. One Artilect will be worth the lives of billions of people. Terrans will oppose their construction and will identify with Earth, *Terra*. A billion humans are worth more than one Artilect. These two groups will be bitterly opposed and will eventually come to blows. Cyborgs will remain an ineffective intermediary.

The Artilect scenario fails at one point. Cyborgs are irrelevant because they already Cosmists in theory. Obviously, they also hold to an evolutionary perspective and will seek to advance themselves to Artilect status, despite their fixed station, forever frozen as part machine and part humanoid. Cyborgs will be without genitals and unable to reproduce. Sex is a distraction to evolutionary advance. Godhood or angelic status is sexless. Angels do not procreate, "they neither marry nor are given in marriage, but they are like the angels in heaven."[453] There will be two hostile parties in the future, Earthlings and God-like men and their machines. There is only one hive-mind, just as there will be only one Artilect that rules, so there is only one computer system. Once plugged in economic necessity will make it impossible to disenfranchise yourself from the system. I reiterate, *what will happen to all those who remain unplugged?*

Future Scenarios

Foretell to us what it is that shall happen . . . declare to us things to come!
Isaiah 41: 22
(NAB).

The singularity will leave billions left behind to face a world in turmoil and tribulation without the enhancement of higher consciousness. Singularity is an eschatological belief for the technological society. Millennialism laid the groundwork, Teilhard's Omega Point gives it direction as a materialist version

---

[452] Hugo de Garis, *The Artilect War* (Palm Springs, CA: Etc, 2005).
[453] Mark 12: 25 (NAB).

of the kingdom of God. The City of Man rises over the horizon turning Augustine's City of God on its head. The science fiction writer Robert Sawyer outlined three possible future scenarios derived from popular science fiction movies: termination, subservience or assimilation. First, when the computer achieves self-awareness it will immediately recognize humanity as a threat and seek to annihilate us. Second, in the subservient model, epitomized by the *Matrix Trilogy*, the Artilect does not seek to destroy humanity but to enslave us by reducing people to chattel. Humans are a source for power, property and wealth. We exist to serve the machine. Lastly, the movies give us the cyborg image. Assimilation to the machine captured in *Star Trek: The Next Generation* demonstrates the frightening prospect of becoming one with the Borg. All life forms are incorporated into the cybernetic organism. The Borg is not a dedicated computer, but a hybrid formed from flesh and machine. Sawyer suggested a fourth symbiotic alternative by arguing the emerging AI will not conform to these, "killer expectations."[454] Both humanity and machine will prosper through mutual dependence. He suggested that when the internet becomes self-aware it will not perceive us as a threat but partner in symbiotic relationship. This will lead to cooperation and further advance not eradication, subservience or assimilation. Cooperation with machine intelligence seems quaint and optimistic. Evolutionary forces that produce aggression are generally directed towards intraspecies violence as Konrad Lorenz (*On Aggression*) suggested. The notion that an alternative intelligence would be amicable toward humanity because it must survive in symbiotic relationship with people neglects the fact that this machine will derive its intelligence from human creators. We have no assurances that it will not see humanity as an intraspecies Darwinian threat. All species exists in a symbiotic relationship with members of its own kind. This fact however, does not curb aggressive behavior that a species may feel towards other members if food, mating or territory is threatened. Sawyer views humanity and machine intelligence as cooperative because we will need each other for survival, but this fact does not stop aggressive behavior, especially if there are members in human society perceived as expandable by those more powerful. If the machine will cooperate with some people that does not mean it will cooperate with all people. We have no way of knowing how this new intelligence will react to people outside its sphere of influence and dependence. Sawyer hypothesizes that the World Wide Web will eventual evolve self-cognition and welcome humanity as a cooperating partner in its efforts to survive, but if this is the case how will it perceive all those billions who do not own a computer?

I fail to see the difference between assimilated and symbiote since both lead invariably to the cyborg condition. These two alternatives appear to be

---

[454] Lawrence Terlizzese, *Killer Computers: Sawyer Revisited*: http://www.utdallas.edu/c4v/killer-computers-sawyer-revisited/ March 7, 2011.

the same destiny. Assimilation means absorption of one entity by another. It's not equalitarianism. Mortals will be left to fend for themselves in unimaginably worse conditions. All the problems created by technosecular environment will remain such as over population, unemployment, ecological meltdown, rampant drug addiction, crime and war on a colossal scale. The uninitiated will be without any enhancements possessed by the Ascended, who will live in guarded communities, domed cities, luxury sky-scrappers, even the space station, protected by state-of-the-art weaponry giving them an air of invincibility, the most effective deterrent, just as nuclear weapons operate today for the developed nations against the hordes of the East.

Youth and Media

The Future is Electric Youth[455]

It only takes one generation born into a new social environment to accept its conditions as normal, as if the world had always been that way. For the present Millennial generation, there never was a time when the computer was not. They are the first generation raised by the computer and will be the generation in power by the middle of this century and thus directly responsible for the Singularity. This makes them the most pivotal generation in history. We have a measure of control over the future by the values we impart today. Media has replaced traditional institutions such as church, school, family and peers as primary source for values. Media has a greater impact on how youth view what is normal than any of the normalizing agencies. Parents have lost control of their children's future through media overexposure. If traditional socializing agents wish to regain control, they must eliminate media in raising children and curb their own habits. We should do the opposite of whatever TV says, and TV says one thing: watch me. The off button is media technology's greatest invention; it is useless to change the channel to a more appropriate program because visual impact on the brain is the same. This is especially true of news broadcasts the audience does not learn to think through subtle indoctrination, the inception of fear and stifling imagination. Media tell us what to do, it leaves no choices open. This is especially loathsome in political advertising. Elections are dictated by propaganda campaigns not rational debate. The same advertising methods used in commercials that incite people to consume are deliberately inculcated to motivate them to vote. The only defense is the off button. I define propaganda as cessation of dialogue, a societal wide condition focused on one exclusive opinion. Everyone says the same thing. The majority are convinced that the current direction in, "technological progress" is inevitable, irreversible and for the good of all.

---

[455] Debbie Gibson, *Electric Youth*, "Electric Youth" (Atlantic, 1989).

The only way people will accept the cyborg order that leaves the unenlightened world behind is through social conditioning that presents the Singularity as inevitable. All religious, ethnic and national ties to the past must be eradicated so that people do not see any differences. They are socialized to view themselves as all the same. This is the bedrock for a totalitarian system to construct its utopian paradise. "The whole world was ordered according to one language and one mind."[456]

The value of radical uniformity is gaining universal acceptance. Stating simply *all is one.* Ideological, racial, religious and gender distinctions mean nothing more than name calling, and intolerance. All things are equal, a frog is no different than a prince, if you love long enough, you can transform it into any image you wish. "Man, the undetermined animal"[457] recedes into nature through acceleration as one kind of intelligent species, but not something special or worth preserving. Mankind is a concept that will be abolished.

The Computer as Cartoon for Living

We Learn through Play.

The computer is a cartoon for living. Play is the greatest form of learning. To ingrain youth with values, make a game of it. Play presents learning as fun, normal and entertaining. "The great archetypal activities of human society are all permeated with play from the start."[458] Millennials derive fun, livelihood and even partners from video games, apps and websites. They cannot imagine life without the computer. As an extension of themselves it is a defining element to their identity. Innovation is play. Nevertheless, an inherent problem exists to this learning approach. Scientific values and education are not accessible to the average student. Engineering, math, physics, not to mention theology, demands arduous learning habits that can hardly be comparable to fun. There is nothing fun about earning a PhD, writing a book or learning a second language. There is joy at the end of the process, but no immediate gratification normally associated with games. The hard discipline of science needs to perpetuate its culture for technological acceleration to continue. When those values are degraded the infrastructure created around them, eventually collapses. Gasset argued that people want to enjoy the benefits technology gives but grow disillusioned with its discipline necessary to make this advantage accessible ultimately this is recipe for a disaster. Spengler also maintained that when engineers forsake science for

---

[456] Genesis 11: 1.

[457] Nietzsche quoted in Jaspers, *The Future of Mankind*, 194.

[458] Johan Huizinga, *Homo Ludens: A Study of the Play Element in Culture* (Boston: Beacon, 1950), 4.

their salvation then technology will become demonized precipitating decline.[459]

For the technosecular order to survive, it must control the transmission of values by dominating the flow of knowledge. The computer changes the nature of learning by making the book unstable and amorphous. Those raised on the computer were conditioned differently. Knowledge is now an interconnected web, rather than an ascending staircase. Logic is not absolute, but malleable to needs. Only one thing is certain, acceleration into a predetermined future. Technology conditions young people to expect learning to be easy and effortless. This leads youth to avoid the more difficult and abstract. In focusing learning in one area, the technical expert is learned, but ignorant outside his expertise. Specialization in knowledge results in what Gasset called, "the learned ignoramus"[460]; it is also cause for the cult of amateur. This condition implies a wistful familiarity with general knowledge allowing the inductee a false assurance, a jack of all trades and master of none. The leveling process or equalization of values afforded by media technology has created ripe conditions in the current generation for the disappearance of humanity as a viable concept in its next evolutionary jump: The Singularity.

Superman is not a Cyborg

To sin against the earth is now the most dreadful thing.
Thus Spoke Zarathustra

In his doctrine on the superman, Nietzsche made his signature injunction: "Man is something that shall be overcome."[461] However, this must not be mistaken for anticipating a cyborg race, but Renaissance for the human spirit. His goal was not to replace modern man with a contrivance, but to reintegrate humanity into nature. He hardly gives a ringing endorsement to the clock work universe. "Let us even beware of believing that the universe is a machine: it certainly was not constructed for one purpose and calling it a 'machine' does it far too much honor."[462] Mankind is Lord of Earth. And the superman was the renewal of Earth through our return to a natural milieu. Speaking through his fictional mouthpiece Zarathustra, he implored us to avoid those wanting to transcend the Earth. "The overman is the meaning of the earth . . . I beseech you . . . remain faithful to the earth and do not believe those who speak to you of other worldly hopes! Poison-mixers are they . . . despisers of life . . . decaying and poisoned . . . whom the earth is weary: So

---

[459] Gasset, *Revolt of the Masses*; Oswald Spengler, *The Decline of the West*, 2vols (New York: Knopf, 1938).

[460] Gasset, *Revolt of the Masses*, 112.

[461] Fredrich Nietzsche, *Thus Spoke Zarathustra: A Book for None and All* trans., by Walter Kaufmann (New York: Penguin, 1966 [1891], 12.

[462] Idem, *The Gay Science* trans., by Walter Kaufmann (New York: Vintage, 1974 [1887]), 167 {109}.

let them go . . . To sin against the earth is now the most dreadful thing, and to esteem the entrails of the unknowable higher than the meaning of the earth."[463]

The, "entrails of the unknowable" is to lust after forbidden fruit. When knowledge is pursued against life it is self-defeating, "Now, is life to rule over knowledge, over science or knowledge to rule over life? Which two authorities is the higher and decisive one?" Nietzsche inquired, "No one will doubt: life is the higher, the ruling authority, for any knowledge which destroys life would also have destroyed itself. Knowledge presupposes life and so has the same interest in the preservation of life which every being has in its own continuing existence. Thus, science requires a higher supervision and guarding."[464] Regrettably, modern science is still a relatively immature discipline in need of guidance from the older established fields, such as philosophy and theology. Science is murderous and misanthropic in its very nature; it kills its subjects. Science did not hesitate to commit patricide in rejecting metaphysics and matricide in casting off theology as a sound basis for knowledge and adopted a truncated epistemology. The, "scientific method" shamelessly flaunts itself as, "objective" despite the fact that it measures only material reality; rejecting classical dualism between spirit and matter, which served as intellectual framework for the Western world. Modern science does not recognize the existence of the ideal, the spirit, but flatly denies its validity, thereby enslaving itself to monist cause and effect determinism.

Nietzsche argued that the scientific pursuit of knowledge was pretext against the forfeit to life its method has cost the modern soul. "Science . . . lives in inner contradiction with eternalizing power of art and religion," he said, "so far as it hates forgetting the death of knowledge, so far as it seeks to remove all horizon-limitations and throws man into an endless—unlimited light-wave-sea of unknown becoming."[465] He thought that science was an escape from the tragic sense of life, a sly buffer to protect from the plain truth. Social arrangements, "the conditions of life" or, "power relations"[466] may include error, so that even our contemporary technosecular society must never succumb to total and final order, since it is not inerrant and unimpeachable. Nietzsche posed an acute dilemma for the unbounded greed for knowledge and power, a quest that has proven truly Faustian. What if knowledge turns out to be destructive, antithetical to life? Do we continue in

---

[463] Nietzsche, *Thus Spoke Zarathustra,* 13.

[464] Idem, *On the Advantage and Disadvantage of History for Life* trans., by Peter Preuss (Indianapolis: Hackett, 1980 [1874]), 62.

[465] Ibid.

[466] Idem, *The Gay Science*, 177 {121}; Michel Foucault, *The Foucault Reader*, Ed., Paul Rainbow (New York: Pantheon, 1984).

this knowledge and die or live free without it? The answer is self-evident, knowledge which destroys life destroys itself knowledge presupposes life. Without life there is no knowledge.

The Ultimate Sin

The Pharisees . . . said, "This man drives out demons by the power of Beelzebub,
the prince of demons"
. . . every sin and blasphemy will be forgiven,
but blasphemy against the Holy Spirit will not be forgiven.
Matthew 12: 24, 31
(NAB)

Scientific naturalism has proven destructive of the material world, despite its materialism. The dialectic between spirit and matter has broken down with disastrous consequences. The zeal for objectivity in science has rejected any subjectivity as a measure for truth. "The followers of Galileo unfortunately fended off any inquiry into their own subjectivity," Mumford assured us, "in rejecting values, purposes and non-scientific meanings, fantasies, dreams [subjectivity] as irrelevant to their positive methodology, they failed to recognize the part such subjectivity had played in creating their own system."[467] The end result was absolutizing knowledge for its own sake and the displacement of human perfection as its guiding principle. They had eliminated every value and higher purpose, but their own pursuit of scientific truth, which became the one absolute their universe revolved around. This represented the infinite expanse of knowledge, which was their only driving categorical imperative. Their one unconditional moral obligation that was binding at all times.

Knowledge is power represents the absolute operative principle. Theologian Karl Rahner defined absolute as, "That which exists in and of itself, altogether exempt of any dependence."[468] This is also the definition for God's aseity (self-existence), God's eternal nature whereby he exists in-of-by himself. The analogous comparison between science and God leads to the invariable conclusion that for its adherents *science is God.* The time for choices has come. Do we seek the god of knowledge or the all-knowing God? Do we follow, as it has been called, the "devil's principle: what can be done must be done?"[469] Edward Teller related this conviction when he asserted, "there is no case where ignorance should be preferred to knowledge—especially if that knowledge is terrible."[470] Life must be the limit to our knowledge. Any

---

[467] Lewis Mumford, *The Pentagon of Power: The Myth of the Machine* Volume Two (New York: Harvest, 1970), 74.

[468] Karl Rahner, *Dictionary of Theology* New Revised Edition (New York: Crossroad, 1981), 1.

[469] Theodore Ziolkowski, *The Sin of Knowledge: Ancient Themes and Modern Variations* (Princeton: Princeton University Press, 2000), xii.

[470] Edward Teller quoted in Ziolkowski, *The Sin of Knowledge,* xii.

knowledge, such as nuclear weapons which threatens life and planetary survival is anti-life and anti-earth. Jonathan Schell writes concerning nuclear weapons, "They grew out of history, yet they threaten to end history. They were made by men, yet they threaten to annihilate man. They are a pit into which the whole world can fall—a nemesis of all human intentions, actions, and hopes. Only life itself, which they threaten to swallow up, can give the measure of their significance."[471] Our destiny is being suggested toward two different futures. One is forbidden and the other free. The choice is clear the fear of the Lord is the beginning of knowledge. Choose Life!

Nietzsche had underestimated the depth of madness the lure of science had on us. Like a glow worm, the enticement to further knowledge, the temptation to power proves irresistible. The modern sin is to pursue the unknown at the expense of human life and the survival of the Earth itself. This is *the ultimate sin*, blasphemy against the Holy Spirit, the unpardonable sin unto death.[472] This is transvaluation attributing good to evil, evil to good, just as the Pharisees ascribed the work of Christ to demons.[473] Technology that leads to death is apostasy. Science since the invention of the atom bomb has proven itself anti-human, and necessarily anti-Christ. "The physicists who so steadfastly pursued the splitting of the atom were in fact," Mumford reminded us, "endangering the human race. With the invention of nuclear bombs, they placed all life on earth in danger."[474] Oppenheimer the father of the atom bomb confessed, "We have done the devil's work."[475] Rightly spoken did Nietzsche say, "There is a great deal I do not want to know."[476] Wisdom must limit knowledge, however, terrible that knowledge might be. There is no sense in pursuing technology that will ultimately destroy us and the world itself that is the true terribleness of the scientific project.

---

[471] Jonathan Schell, *The Fate of the Earth* (New York: Avon, 1982), 3.

[472] First John 5: 16, 17.

[473] Matthew 12: 22-37; Abraham Kuyper, *The Work of the Holy Spirit* trans., by Henri De Vries (London, UK: Funk & Wagnalls, Co., 1900), 608-612.

[474] Mumford, *The Pentagon of Power,* 75.

[475] J. R. Oppenheimer quoted in Jaspers, *The Future of Mankind*, 199.

[476] Fredrich Nietzsche, *Twilight of the Idols/Anti-Christ* trans., by R.J. Hollingdale (New York: Penguin, 1968 [1888]), 33.

CHAPTER TEN The Victory of Hope

I Am Doing Something New.
Isaiah 43: 19
(NJB)

Newness belongs exclusively to the work of Jesus Christ, "I make all things new," he said. This is the Eternal Now, the victory of hope. Hope is only possible on the basis of freedom. A world that remains open to the future is available to the eternal Spirit. They live in the Eternal Now. A world that is closed serves a temporal order bound to its own internal necessity. They accelerate into a forbidden future. One that is entirely predictable because it is empty of freedom. Where there is criticism there is hope; criticism liberates our perspective to a see a different future, free from social constraint and subject to change and therefore victorious. Technology serves the cause of progress in freedom or else technology is not free, but only rapid movement pretending to be progressive. Human extinction is one potential future entirely avoidable once we accept free responsibility for the world we choose to create now. Resistance is freedom!

The Eternal Now

God is New Every Moment.

Awareness of our role in history determines our sense of purpose. Everything until now has brought us to this moment. What happens next is entirely up to us. There are limits to fore casting. Jaspers tells us we can only predict the negative. What is doomed to fail: the fall of a city in war, the death of the sick, the extinction of a species. The spontaneous future is unpredictable. Freedom cannot be programmed, "experience tells us that the impossible can happen that the probable may never happen and above all that new, completely unthought of realities may appear."[477] This is the Eternal Now, the instant that eternity breaks in on us with a purpose oriented toward a future that has never been. A future we chose to create now. The future unveiled in the present by remembering the past. We have excellent grounds for hope, despite all calamity the human spirit triumphs. Resurrection always follows crucifixion. "That I may know him, and the power of his resurrection,

---

[477] Karl Jaspers, *The Future of Mankind* trans., by E. B. Ashton (Chicago: The University of Chicago Press, 1961), 282.

and the fellowship of his sufferings," the Apostle uttered.[478] Death never has the last word. Humanity will triumph over the forces arrayed against it by reason of its own indomitable will-to-victory. As the first computerized generation Millennials are unique. They hold the future in their hands. They are resilient and will determine it however they wish. They have normalized the computer making it an indispensable tool in modern life. Yet, for the digital revolution to succeed and not relapse into tyranny Millennials must sharpen their critical perception if they want to be more than a pawn in the acceleration process. I believe in progress, so I criticize it to improve it. There is no point to criticism if not constructive. Where there is criticism there is hope. Critical participation causes us to reflect on acceleration without the easy conscience affirmation affords optimism.

The technological imperative is the basic faith assumption for our age. Once we remove that cornerstone young people will be left to rely on their own judgment as to the extent they choose to participate in the system. Criticism allows space for freedom, however, small to disagree and assert a different view. Self-destruction invariable follows any uncritical embrace of limitless potential. Only God is limitless. All other claims represent an unwarranted infringement on the infinite. From its start civilization opposed the purposes of God and usurped his authority through concentrating power in the city. "This is just the beginning. Soon they will be able to do anything they imagine, nothing will be impossible for them." [479] This is the human hive—Man as collective unit. On our way up to heaven God comes down to meet us, "Let Us go down there." When we seize God's power, we abandon our governing position as vice-god, a little god on Earth to Father God in heaven. We play God to our own detriment. God will judge our power grab and break the hammer of technology on the anvil of war and calamity. Already twice in one century this has happened, a third is unfathomable.

Fearless engineering is a subtle refusal to recognize boundaries. Technology is not exempt from limits. The better part of wisdom is knowing what lines should be crossed and what should not. Those responsible for the effects of innovation must receive the praise and the blame for technology's triumphs and failures. Transportation also means pollution. Telecommunications equals the death of privacy. American historian Leo Marx related the importance in asking questions through criticism to improve development. "It is useful to ask what the purpose of this technology is. Only by questioning the assumption that innovation represents progress can we begin to judge its worth."[480] Progress is not exclusively technological.

---

[478] Philippians 3: 10 (KJV).

[479] Genesis 11: 6.

[480] Leo Marx, "Does Improved Technology Mean Progress?" in *Technology and the Future*, 6th Ed, edited by Albert H. Teich (New York: St. Martin's Press, 1992), 14.

In the eighteenth-century progress meant collective advance for all humanity as single unit in all areas. The world moves forward as a whole or not at all. Freedom once the guiding light to progress gives meaning to everything else. Technology served the ends for a radical transformation. Political liberation lost its central status usurped by acceleration. The goal in human perfectibility is now measured by technological complexity not the value of freedom. It was believed that as we became technologically proficient the world would automatically move forward to a better condition. Technology would bring with it the necessary moral and spiritual growth to direct it to ultimate good. Innate to innovation was the wisdom to use it, a moral crisis attributable to new technology was thought impossible.

Everyone's A Critic

Wherever the spirit wished to go, there the wheel went.
Ezekiel 1: 20
(NAB)

I seek to affirm progress while retaining a critical spirit guided by a transcendent frame of reference outside the system. This complements the system with a moderate voice acting as Socratic gadfly or conscience that serves as check and balance. I do not wish to return the world to a pre-electric age, but to provide a spiritual will-to-triumph and control over technology to avoid catastrophe and insure humanity's progress and liberation. Critical discussion works as feedback, a necessary principle for the life and maintenance of machines. Without it the system will be unable to regulate itself and explode or shutdown. Technology has no internal ballast to keep it from extremes. There is no inherent spiritual compass attached to your latest upgrade. According to Cybernetic pioneer Norbert Weiner feedback may be defined as, "a method of controlling a system by reasserting its past performance."[481] Criticism is essential to the health of any society. In creating pockets of freedom from technological constraint we allow for play and flexibility that keeps the system from locking up. The critic's task resembles the watchmen in sounding the alarm that awakens the sleeping population to the dangers just above the horizon, so that they are not overtaken, but given adequate time to meet the threat. The critic notices only potential dangers; we can only predict the negative. We are not doomsday prophets, but are given to the safety and success of the city, "seek the welfare of the city where I have sent you into exile, and pray to the Lord on its behalf, for in its welfare you will find your welfare."[482] War and chaos doesn't help anyone. The critic is unpopular because he disturbs the people's apathy he arouses from slumber

---

[481] Norbert Weiner quoted in Rudi Volti, *Society and Technological Change* 4th Ed., (New York: Worth Publishers, 2001), 10.

[482] Jeremiah 29: 7 (NRSV).

and is often accused of being an enemy of the state because he challenges the status quo. Socrates was executed for corrupting the youth of Athens because he taught them to think for themselves. The Hebrew prophets were exiled when they warned of impeding invasion. Despite all the grandeur, there perhaps might be an ugly side to accelerated change, hidden just beneath the surface that necessitates scrutiny, a tiny liver spot on the skin of civilization, if not caught early the cancerous cell could spread to the whole body.

Law Is Social Technology

A state, is called the coldest of all cold monsters. Coldly lieth it also;
and this lie creepeth from its mouth: "I, the state, am the people." It is a lie!
Thus Spoke Zarathustra

Socratic dialogue is a method of inquiry that uses questions and answers to awaken knowledge in the audience so that they will see a different perspective. One that is uniquely their own rather than simply accept a view because it happens to be popular. History is a form of critical feedback I use to contrast present smugness with its own superiority. Simply because something is the latest trend does not make it right. Every younger generation struggles against, "going with the flow" and the mentality that says, "everyone's doing it," especially Millennials who favor uniformity over difference and equalitarianism over individualism. History serves as teacher to our present and prophet to our future. As philosopher George Santayana stated, "those who cannot remember the past are condemned to repeat it."

History teaches that the more technologically complex a civilization becomes the more centralized and controlling its power structure to the loss of solidarity, freedom, rights and personal dignity. The more vulnerable it becomes to catastrophe, plagued by technical error, obsolescence, sabotage and war. Society is prone to suffer from pollution, resource depletion, even bad weather; for example, transportation in an agrarian society, such as in the Old South was little affected by snow or ice. Today a little snow in any major metropolitan city in the South causes alarm, school and business closings, driving comes to a halt and the city becomes immobilized until the temperature reaches above freezing.

Dependencies or addictions such as petroleum and electricity further demonstrate vulnerability. If the flow of oil or electricity is somehow interrupted it could potentially have disastrous effects. Likewise, digital terrorism has revealed itself as another threat. Hackers seek to overthrow the established order by crashing computer servers demonstrating the strength of centralization and its fatal weakness.

The history of technology shows the lockstep progression of technological complexity with centralized government as inevitable necessity. This phenomenon has proven true across all ideological and cultural differences. Lenin declared, "Communism is Soviet power plus electrification."[483] Lewis Mumford called the first river civilizations in Egypt and Mesopotamia, "Mega-Machines" because they exhibited characteristics of centralized power instituted to maintain control over indiscreet homogenized masses. These people were slaves and corvée workers drafted in the service of colossal building projects, such as the Pyramids.[484] In the early modern period political centralization and control grew concomitantly with technological innovation. In the fifteenth century a technological Military Revolution transformed European warfare with the advent of firearms and cannon. Volley fire in coordination with artillery was the most effective use of firearms technology, which necessitated larger standing armies and centralized state authority to raise, maintain and control military largesse. Modern artillery made short work of medieval stone fortresses and walls provoking a technological escalation and arms race in fortified earthen defenses with its own guns and artillery. The cost of war skyrocketed until only the wealthiest European Sovereigns could afford the new technology. Warfare became the exclusive province of centralized Nation-States paying for innovation through tax revenue, mercantilism and conquest. The ranks of modern armies were peopled through universal conscription, avenues of profit and manpower not available to feudal lords. European aristocracy was limited to their fiefdoms for soldiers and limited budgets, which could not compete with a central authority that drew its resources from the entire nation,

> The Military Revolution introduced competition between states and dynamic social mechanisms that relentlessly favored technical development. The centralizing effects-social, political and economic---were akin to those called forth by irrigation agriculture in the great hydraulic civilizations [River Civilizations: Han, Indus, Mesopotamia]. Thus, Europe acquired, like Egypt and China, thousands of years before, the full panoply of civilized institutions. From the fifteenth century onwards, the creation of national armies and royal navies resulted in political centralization as inevitably as had the hydraulic projects of the ancient and medieval East.[485]

Today the trend towards centralization continues relentlessly. Communication, transportation and military technologies provide the state

---

[483] N. Lenin quoted in James E. McClellan III, and Harold Dorn, *Science and Technology in World History*, 2nd ed., (Baltimore, MD: The Johns Hopkins University Press, 2006), 346.

[484] Lewis Mumford, *The Myth of the Machine: Technics and Human Development* (New York: HBJ, 1966).

[485] McClellan and Dorn, *Science and Technology in World History*, 196-97.

with means of extending its might in every geographic locale, never possible until modern times. The Roman Empire could project its will thousands of miles away from Rome, but due to limited technological means the Legions could not present an ubiquitous presence throughout its boarders. Power was concentrated in urban centers, leaving the countryside and larger provinces ungoverned and autonomous. Freedom from Roman tyranny could easily be accomplished through taking flight and refuge in the hinterland, where the Romans could not reach. Technology enlargers the reach of the State so that as technology advances the state grows accordingly. "Call to mind what the state was at the end of the XVIII century . . . Quite a small affair," Gasset intoned.[486] By refusing to recognize any limits to its power the State serves as the Apparat for technology and is subject to the same strangle hold imposed by specialization; enormously successful but paralyzed in transmitting its values to the next generation.

If the Singularity transpires according to the Transhumanists vision it will necessarily embody one global will, capable of solving all, hitherto, unsolvable problems issuing from technological acceleration: overpopulation, pollution, environmental devastation, resource depletion, poverty, inequality and war. All will be rectified through the inescapable one world order imposed by the transcendent intelligence. We create a new digital god in exchange for freedom. God-killers are needed more now than ever, if we wish to remain free people in a future determined by Machinehead. Albert Camus said, "To kill God is to become god oneself; it is to realize already on this earth the eternal life of which the Gospel speaks."[487] Never has Sartre's criticism rung truer than before our technologically created deity, "If God exists, man is not free."[488] If we choose freedom the digital God will never exist.

The Future as Human Extinction

He called us who were not; and was pleased from nothing to give us being.
The Second Epistle of Clement to the Corinthians 1: 10

Desire for new technology has been predetermined by visual media. The futurist industry has hijacked, space-jacked, if you will, modern progress to enshrine the master-slave order into perpetuity. The vision of a future technological cornucopia is largely an advertising gimmick, a clever illusion perpetrated as corporate state propaganda for indoctrination purposes and stimulation of the economy. Futurism as de facto school of thought embodies the technological imperative for the 21st century and appears to be the only

---

486 José Ortega y Gasset, *The Revolt of the Masses* (New York: Norton, 1932), 117.

487 Albert Camus quoted in Vahanian, *The Death of God*, ix.

488 Jean-Paul Sartre, *Being and Nothingness* (New York: Washington Square Press, 1956).

viable intellectual framework for a materialistic age cut asunder from history, philosophy and religion. It will attempt to free the future from Earth's climate and gravity, but not before it drains all resources through reclamation of Earth's natural environment by replacing it with an artificial one.

British Philosopher Bertrand Russell noted simply that modern science and technology has menaced human existence by increasing the boundaries of human knowledge and power over nature, but has not matched improved material conditions with the depth of spiritual character responsible enough not to self-destruct. "Science has not given men more self-control, more kindliness, or more power of discounting their passions," he said, "Men's collective passions are largely evil; far strongest are hatred and rivalry directed towards other groups. Therefore, at present all that gives men power to indulge their collective passions is bad. That is why science threatens to cause the destruction of our civilization."[489] The transvaluation from metaphysics to physics leaves the technological order stranded in rank materialism, which finds humanity's only hope in space exploration, which is a far cry from a humanitarian agenda, but its opposite species suicide.

The success or failure in the endeavor to replace nature and traditional society with an artificial simulation superior to the original, although an obvious forgery, constitutes what Professor Jacques Ellul called *L'enjeu du siècle* (the gamble of the century) in attempt to create a better world than we inherited from our past the technological order will risk the survival of the human race and the planet earth itself. The final dialogue at the close of Things to Come, (1936) a cinematic Manifest Destiny for outer space, we hear the technological credo, "All the universe or nothingness? Which shall it be, Passworthy? Which shall it be?" We are led to believe that if humanity does not advance forward and finish aggressive reclamation in the image of the machine it faces annihilation as consequence of its failure. Ellul stated confidently that this concept of the future as either technological utopia or oblivion is nothing more than a, "gigantic bluff" concealing technological failure and amplifying its success, "when I say bluff, it is because so many successes and exploits are ascribed to techniques (without regard for the cost or utility or risk), because technique is regarded in advance as the only solution to collective problems . . . or individual problems . . . even the meaning of life . . . it is seen as the only chance for progress and development in every society."[490]

Belief in a limitless future in outer space is inherently genocidal and represents an agenda of calculated annihilation. The Technological Imperative

---

[489] *Bertrand Russell, The Future of Science (New York: Wisdom Library, 1959), 53.*

[490] Jacques Ellul, *The Technological Bluff*, trans., by Geoffrey W. Bromiley (Grand Rapids: Eerdmans, 1990), xvi.

is intrinsically permissive of human extinction and reckless in its push to acquire knowledge and destroy all natural and traditional boundaries. The technological imperative means, finally, genocide. Simply, because it has become technically possible to do so with the advent of our technological and therefore political superior: *The Cyborg*. Subjugation to the machine and eventual replacement does not represent technological utopia gone awry, but its logical conclusion. Posthuman nihilism is by far the ugliest and most gruesome anti-humanism from any current or past vision. It poses a destiny for us first as slaves, then, replaced by greater machine intelligence. The dystopian futures depicted in most science fiction movies such as *The Matrix* and *Terminator* franchises are not examples of Luddite hysteria, but represent a perfectly acceptable destiny. Post Darwinism (Artificial Selection), and Transhumanism stoically assent to human extinction as natural and *the inevitable course of things*.

Resistance is Freedom

Vive la résistance!

Determinism is the philosophical context for, "the Borg" the arch nemesis of the universe in *Star Trek: The Next Generation* TV series, which never tires of repeating its asinine mantra: "Resistance is Futile!" Resistance is futile if we believe it is futile. However, because the Borg feels compelled to announce its plan to assume control, it demonstrates that *resistance is the path to freedom*. If resistance was futile there would be no need to tell everyone. The Borg would simply assume control, and everyone will wake up in a dream that they are in the mind of the computer without ever knowing anything different. Assimilation, however, cannot be coerced, the seizure of power results in eradication of the human will not assimilation. Effective assimilation results from two separate entities mutually agreeing to surrender their wills, likes and differences one to the other in hope of *common destiny*, not unlike the *One Flesh* union.[491] All potential inductees into the machine must be persuaded to surrender by convincing them that they have no other choice. Passivity is most effective through intimidation. It immediately overwhelms the senses, *especially sight* with colossal scale that darkens faith's horizon with a monstrous megastructure in the shape of a perfect cube that *appears* to be both the *immovable object* and the *irresistible force.* Psychological warfare is essential to the machine's success at subduing humanity. The machine reveals its impassible character by resorting to fear. "People will faint from terror, apprehensive of what is coming on the world," Jesus Said.[492] The majority of people are easily motivated by fear. Immediate submission is primary achieving *totality.* Slaves must be conditioned to servitude by imprinting upon them that *slavery is destiny.* Happy slaves do not know they are slaves and

---

[491] Genesis 2: 24; Matthew 19; First Corinthians 6: 12-20.

[492] Luke 21: 26 (NIV).

accept the master-slave social arrangement as unalterable fate. Human nature will grow to love whatever it must do or is forced to do if it knows no alternative and is denied all other choices. Thus, *resistance is freedom*!

What Should I Do?

You know what to do! Freedom cannot be dictated. It is useless to tell people what to do. The question itself reveals a basic dishonesty. On one hand it shows a genuine concern on the other it expresses subtle disdain. "What should I do?" Really means, tell me what to do. It says, "I want to appear to care, but I do not want to change. I will go through the motions without getting involved. I remain decidedly uncommitted." The inquisitor wishes to continue in emotional ambivalence because it spares him the challenge to decide. He waits comfortable with superficial thoughts, a variety of amusements and a host of electric distractions. Ignorance is bliss; it leaves him oblivious and unattached. Listen to the still small voice heard only in quiet reflection.

Where there is criticism there is hope. Neither despair is right when certain, nor optimism when cowardliness asserts everything will be alright. Without a change to move us forward, we stay perpetually distracted and numb to the fate of our future. We will be detached from action and harden to what remains undone by refusing to believe that every choice, every deed, every little word has enormous consequences, now, and for the future of mankind. Every moment is the right time to act every person is responsible to change the world by changing herself. What cannot be done alone, we can do together. What we create we must redeem only then can we control it. If we do not awaken to the danger technology presents—our fate is sealed. We have no one to blame but ourselves. Which shall it be Passworthy, Everything or Nothing?

Conclusion

The Devices of the Wicked Man Shall be Scrutinized.
Wisdom 1: 9
(NAB)

Imperatives are a common ideal or belief. They represent an objective expression of some ultimate concern. Imperatives are social movements guided by belief in irrevocable destiny. They leap into the void of infinite space by reason of primordial impulse to die or kill for one's cause, to tempt death for eternal glory. Imperatives punctuate history with mass suicide and genocide in the hope for a better world to come; examples include the Fall of Jerusalem and Masada, the Crusades, Manifest Destiny, Custer's Last Stand, the Final Solution, Kamikaze, Jihad, Hiroshima, Jonestown. Imperatives are not limited to radical religious sects, but usually embrace the whole society. They claim divine authority to excuse their behavior, which is no less true for secular causes as it is for religious ones. They are intrinsically idolatrous, operating as ultimate directives for the powers that be and invariably compete with the creator. They express the endgame with disastrous consequences. Imperatives paralyze freedom with a fixed cause that allows for no other choices. There is no freedom when only one possibility exists. This is untrammeled necessity and pure determinism. The technological imperative serves this purpose in our technosecular world, exalting the measurable as the only reality and relegating the ideal to fantasy. Adherents have been seduced by their own greed and lust for power. Imperatives debunk all reality as illusory, except its own. They are inherently totalitarian and sway all for the cause. No quarter is given. They will cross every boundary to achieve their aim. Imperatives are fueled by delusion and a false sense of assurance that they are the epicenter of all things. They transform ultimacy into its opposite, the usurpation of spirit with pursuit of material and political causes. And like the Cyclopes it totters under its own weight and lack of perspective. The individual can stand against a social imperative by witnessing to its idolatry and pointing to a higher and genuine concern. We call all to regeneration and not assassination; a mission for which we need a power greater than any technology.

Any liberating mission must begin with the renunciation of the technological imperative. This involves curbing greed and curtailing techno pride, which is nothing more than ambition to exceed known limits coupled with lust, a desire to go beyond given boundaries. In that mission we declare the peril of the hyperreal located directly in the nerve center of the idol: media technology. Machinehead is a Madison Avenue production prompting mindless consumerism with promises of a better tomorrow through instant gratification. Iconoclasm, then, is our calling. Iconoclasm is the destruction of idols; the realignment of faith in the creator. Faith includes a passionate belief in truth and a radical disbelief in the lie that human beings can serve as their own god without self-destruction.

Iconoclasm is not the deconstruction of all images that would be Nihilism, but the desecration of false images that pretend to Godhood. It means the desacralization of the secular. God is not the sum total of our imagination. By creating God in our own image, we give moral sanction to our best values, which is, more often than not, a product of our environment. In technosecular society, God is not an ideal man he is just the image of one. At the current time technology competes with the Creator, for our attention, adoration and faith. The image of God in the form of God, Jesus Christ, remains our measure. Transhumanism attacks the body as the Temple of God by rendering it obsolete through technological simulation. The body glorious will receive full emancipation only through resurrection, which cannot be wrought by human hands.

Epilogue

Throwaway each of you the detestable things that have held your eyes.
Ezekiel 20: 7
(NAB)

Technological convergence is not progress, but manifest regression. If we will not control our cellphone use, we are condemned to upgrade, how can we expect governments to abolish nuclear weapons or stop school shootings? Impotence at home, in the local sphere, suggests paralyses at federal and corporate levels. Managing our technology should not be any more difficult than exercising self-control. But this is where the problem really begins. We have come to rely too much upon our devices. We have lost self-discipline. A push button world leaves no room for impulse control. We indulge ourselves with each use. We are rewarded by an emotional rush. A slight chemical stimulation over time through repeated use becomes addiction. Once the body adjusts to the added stimulus, it can't go without it. I saw this first hand when teaching college rhetoric. Students were given a simple voluntary assignment, abstain from cellphones for 48 hours and write about the experience. Half of them could not finish and experienced marked withdrawal symptoms. They became angry and anxious. They felt lonely and isolated. Too much of a good thing is always bad. Only in the absence of the idol do we reveal our degrading dependence on it. We have lost our autonomy, freedom and passion to live in any other way than how the idol directs. I recommend short periods of abstinence from any dependency to release its hold. This is a great place to start if you are still in denial over your addiction.

More than this we need not add; let the last word be, he is all in all!
Sirach 43: 28
(NAB)

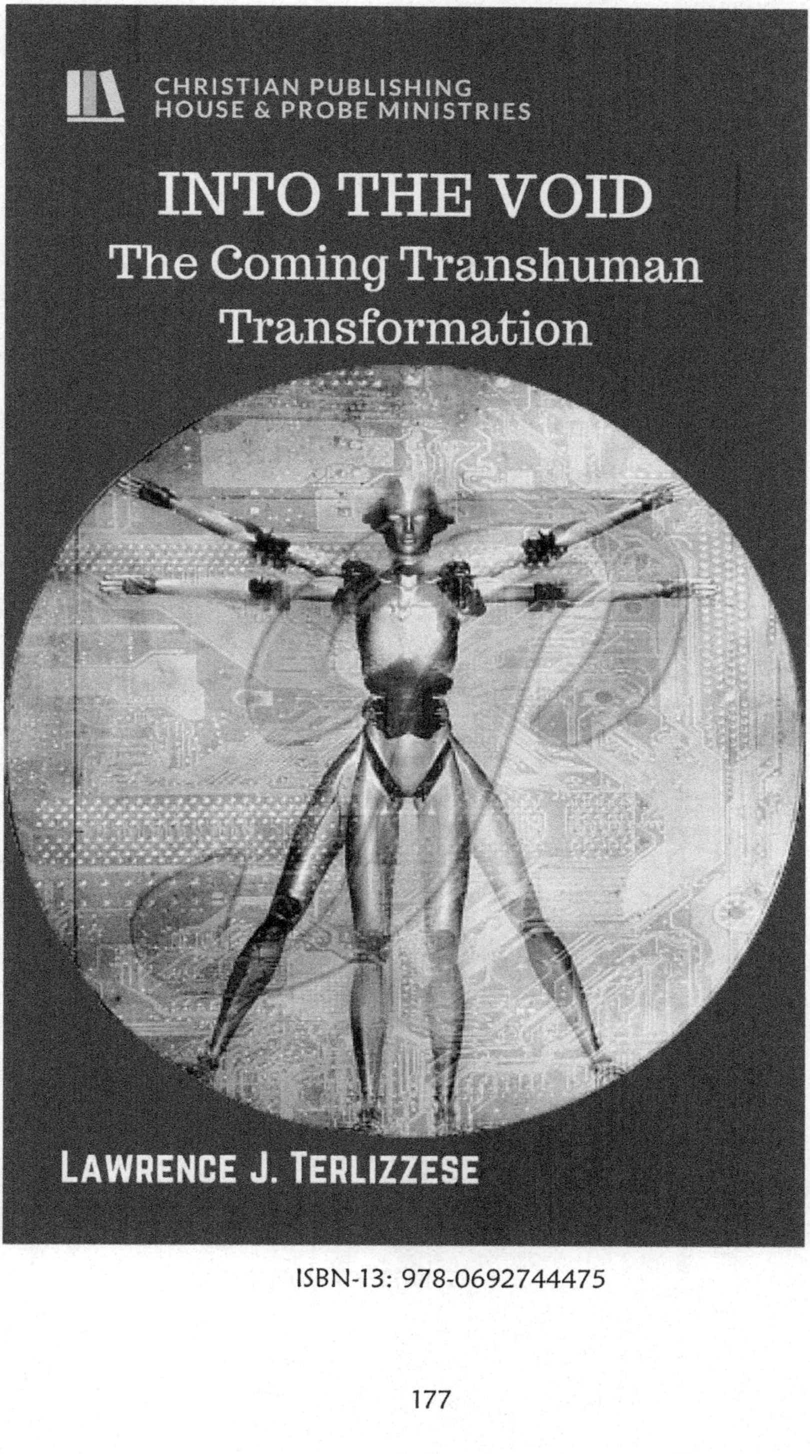

ISBN-13: 978-0692744475

THE ENLIGHTENMENT PRINCIPLE OF SAPERE AUDE (DARE TO THINK FOR YOURSELF) IS BEING FORGOTTEN IN FAVOR AN ARTIFICIAL INTELLIGENCE THAT DOES ALL OUR THINKING FOR US

KILLER COMPUTERS

SCIENCE FICTION ANTICIPATES OUR FUTURE

LAWRENCE J. TERLIZZESE

ISBN-13: 9781071496329

BIBLIOGRAPHY

Adas, Michael *Machines as the Measure of Men: Science, Technology, and Ideologies of Western Dominance* (Ithaca, NY: Cornell University Press, 1989).

Bailey, Lee Worth *The Enchantments of Technology* (Chicago: University of Illinois Press, 2005).

Barbour, Ian *Science and Secularity: The Ethics of Technology* (New York: Harper and Row, 1970).

Baudrillard, Jean *Simulacra and Simulation* trans., by Shelia Faria Glaser (Ann Arbor, MI: The University of Michigan Press, 1994 [Éditions Galilée, 1981]).

Benz, Ernst *Evolution and Christian Hope: Man's Concept of the Future from Early Fathers to Teilhard de Chardin*, trans., by Heinz G. Frank (New York: Doubleday, 1966).

Berdyaev, Nikolas *The Realm of Spirit and the Realm of Caesar* (New York: Harper, 1952).

__________., *Slavery and Freedom* trans., by R. M. French (New York: Charles Scribner's Sons, 1944).

__________., *Toward a New Epoch* (London, UK: Bles, 1949).

Booker, M. Keith *Alternate Americas: Science Fiction Film and American Culture* (Westport, CT: Prager, 2006).

Borg, Marcus *Jesus and Buddha: The Parallel Sayings*, Ed., (Berkeley, CA: Ulysses, 1997).

Braaten, Carl E. *The Future of God: The Revolutionary Dynamics of Hope* (New York: Harper and Row, 1969).

Boyer, Paul *By the Bomb's Early Light* (New York: Pantheon, 1985).

Bradbury, Ray *Fahrenheit 451* (New York: Ballantine, 1953).

Buber, Martin *I and Thou* 2nd Edition. trans., by Ronald Gregor Smith (New York: Charles Scribner's Sons, 1958).

Bultmann, Rudolph *Jesus Christ and Mythology* (New York: Scribner's 1958).

Burtt, E. A. *The Metaphysical Foundations of Modern Science* (Garden City, NY: Doubleday, 1954).

Bury, J. B. *The Idea of Progress: An Inquiry into its Growth and Origins* (New York: Dover, 1932).

Calvin, John *Institutes of the Christian Religion*, Volume One trans., by Ford Lewis Battles (Philadelphia: The Westminster Press, 1960 [1559]).

Canto, Christophe and Odile Faliu, *The History of the Future: Images of the 21st Century* trans., Francis Cowper (Paris: Flammarion, 1993).

Cassirer, Ernst *et al.*, Eds., *The Renaissance Philosophy of Man* (Chicago: The University of Chicago Press, 1948).

Cave, Stephan *Immortality: The Quest to Live Forever and How it Drives Civilization* (New York: Crown, 2012).

Clark, Andy *Natural-Born Cyborgs: Minds, Technologies and the Future of Human Intelligence* (Oxford University Press, 2003).

Clarke, Arthur C. *The Promise of Space* (New York: Bantam 1968).

__________., *Profiles of the Future: An Enquiry into the Limits of the Possible* (n.p., 1973).

Clarke, I. F. *The Pattern of Expectation: 1644-2001* (New York: Basic Books, 1979).

Comte, Auguste *A General View of Positivism*, Centenary Edition, trans., by J. H. Bridges (New York: Robert Speller & Sons, Publishers, 1975).

Corn, Joseph J. and Brian Horrigan, *Yesterday's Tomorrows: Past Visions of the American Future* (Baltimore: The John Hopkins University Press, 1996).

Cornford, F. M. *From Religion to Philosophy: A Study in the Origins of Western Speculation* (New York: Harper & Row, 1957).

Cornforth, Maurice *Materialism and the Dialectical Method*, 4th Edition. (New York: International Publishers, 1972).

Cousins, Norman *Modern Man Is Obsolete* (New York: Viking, 1945).

Davis, Devra *Disconnect: The Truth About Cellphone Radiation* (New York: Writer's House 2013).

Darren Tofts, *et al.* Eds. *Prefiguring Cyberculture: An Intellectual History*, Ed. (Cambridge, MA: MIT Press, 2003).

Degroot, Gerard J. *Dark Side of the Moon: The Magnificent Madness of the American Lunar Quest* (New York: New York University Press, 2006).

Dickens, Charles A Christmas Carol: by Charles Dickens; illustrated by Roberto Innocenti (Italy, 1990[1843]).

Dickerson, Matthew *The Mind and the Machine: What it Means to be Human and Why it Matters* (Grand Rapids, MI: Brazos, 2011).

Dinello, Daniel *Technophobia! Science Fiction Visions of Posthuman Technology* (Austin: University of Texas Press, 2005).

Umberto Eco, *Travels in Hyperreality* (New York: HBJ, 1986).

Ellul, Jacques *The Technological Bluff* trans., by Geoffrey W. Bromiley (Grand Rapids: Eerdmans, 1990).

__________., *The Technological Society,* trans., by John Wilkinson (New York: Vintage, 1964).

__________., *Hope in Time of Abandonment* trans., by C. Edward Hopkins (New York: Seabury 1973).

__________., *The New Demons,* trans., by C. Edward Hopkin (New York: Seabury, 1975).

__________., *The Betrayal of the West* trans., by Matthew J. O'Connell (New York: Seabury Press, 1978).

__________., *Propaganda: The Formations of Men's Attitudes* (New York: Vintage, 1965).

__________., *The Humiliation of the Word* (Grand Rapids: Eerdmans, 1985).

__________., *Perspectives On Our Age* (New York: Seabury, 1981).

Ferkiss, Victor *Technological Man: The Myth and the Reality* (New York: Mentor, 1969).

__________., *The Future of Technological Civilization* (New York: George Braziller, 1974).

Ferré, Frederick *Philosophy of Technology* (Englewood Cliffs, NJ: Prentice Hall, 1988).

Feuerbach, Ludwig *The Essence of Christianity* trans., by Marian Evans (New York: Calvin Blanchard, 1855 [1841]).

Flew, Anthony *A Dictionary of Philosophy* Revised Second Edition (New York: St. Martin's Press, 1984).

Florman, Samuel C. *Blaming Technology: The Irrational Search for Scapegoats* (New York: St. Martin's Press, 1981).

Frank, Allan *The Science Fiction and Fantasy Film Handbook* (Totowa, NJ: Barnes and Noble, 1982).

Fromm, Erich *Ye Shall Be As Gods: A Radical Interpretation of the Old Testament and its Tradition* (New York: Fawcett, 1969).

__________., *The Sane Society* (New York: Premier, 1965).

Fuller, R. Buckminster *Utopia or Oblivion: The Prospects for Humanity* (New York: Bantam, 1969).

__________., *Operating Manuel For Spaceship Earth* (Zurich, CH: Lars Müller, 2015 [1969]).

Gasset, José Ortega y *The Revolt of the Masses* (New York: Norton, 1932).

Garis, Hugo de *The Artilect War: Cosmists vs. Terrans: A Bitter Controversy Concerning Whether Humanity Should Build Godlike Massively Intelligent Machines* (Palm Springs, ETC., 1995).

Haeckel, Ernst *The Riddle of the Universe at the Close of the Nineteenth Century* trans., by Joseph McCabe (New York: Prometheus, 1992 [1900]).

Halacy, D. S. Jr., *Cyborg: Evolution of the Superman* (New York: Harper & Row, 1965).

Harai, Yuval Noah *Homo Deus: A Brief History of Tomorrow* (New York: Harper, 2017).

Harper, Kyle *The Fate of Rome: Climate, Disease and the End of an Empire* (Princeton, NJ: Princeton University Press, 2017).

Hegel, G. W. F. *The Phenomenology of Mind*, trans., by J.B. Baillie (New York: Harper, 1967 [1807]).

__________., *Reason in History: A General Introduction to the Philosophy of History* trans., by Robert S. Hartman (Indianapolis: Bobbs-Merrill, 1953).

__________., *The Philosophy of History* trans., by J. Sibree (Buffalo, NY: Prometheus, 1991).

Heidegger, Martin *The Question Concerning Technology and Other Essays* trans. By William Lovitt (New York: Harper, 1977).

__________., *Being and Time* (San Francisco: Harper, 1962).

Hellman, Louis *Architecture for Beginners* (New York: Writers and Readers Publishing, 1988).

Herbert, Frank *Dune* (New York: Ace, 1965).

Herman, Arthur *The Idea of Decline in Western History* (New York: Free Press, 1997).

Hill, Stephen *The Tragedy of Technology: Human Liberation verses Domination in the Late Twentieth Century* (London, UK: Pluto Press, 1988).

Hodge, Charles *Systematic Theology*, Volume One (Grand Rapids: Eerdmans, 1985 [1871]).

Hordern, William E. *A Layman's Guide to Protestant Theology*, Revised and Expanded Edition (New York: MacMillan, 1968).

Huntington, Samuel P. *The Clash of Civilization and the Remaking of World Order* (New York: Simon & Schuster, 1996).

Huizinga, Johan *Homo Ludens: A Study of the Play Element in Culture* (Boston: Beacon, 1950).

Huxley, Aldous *Brave New World Revisited* (New York: Perennial, 1965).

__________., *Brave New World* (New York: Random House, 1932).

Jaspers, Karl *The Future of Mankind* trans., by E. B. Ashton (Chicago: The University of Chicago Press, 1961).

__________., *Way to Wisdom: An Introduction to Philosophy* trans., by Ralph Manheim (New Haven, CT: Yale University Press, 1954).

__________., *Man* in *the Modern Age* (New York: Anchor, 1951).

James, Simon J. *Maps of Utopia: H. G. Wells, Modernity & the End of Culture* (New York: Oxford University Press, 2012).

Johnson, William, Ed. *Focus on The Science Fiction Film* (Englewood Cliffs, NJ: Prentice-Hall, 1972).

Jones, D. F. *Colossus* (New York: G.P. Putman's Sons, 1966).

__________., *The Fall of Colossus* (New York: G.P. Putman's Sons, 1974).

Jünger, Friedrich Georg *The Failure of Technology* (New York: Gateway, 1949).

Kahler, Erich *Man the Measure: A New Approach to History* (New York: Meridian, 1943).

Kierkegaard, Soren *The Point of View* (London: Oxford University Press, 1939).

__________., *Concluding Unscientific Postscripts* trans., David Swenson and Walter Lowrie (Princeton: Princeton University Press, 1941).

__________., *The Present Age and of the Difference Between a Genius and an Apostle* trans. Alexander Dru (New York: Harper, 1940).

Kirk, Russell *Beyond the Dreams of Avarice: Essays of a Social Critic* Rev. Ed., (Sherwood, Sugden & Co., Peru: IL, 1991 [1956]).

Kuyper, Abraham *The Work of the Holy Spirit* trans., by Henri De Vries (London, UK: Funk & Wagnalls, Co., 1900).

Lewis, C.S. *The Abolition of Man* (New York: Macmillan, 1947).

Locke, John L. *The Devoicing of Society: Why We Don't Talk To Each Other Anymore* (New York: Simeon and Schuster, 1998).

Mackay, Charles *Extraordinary Popular Delusions and the Madness of Crowds* (Wordsworth: Denmark, 1995 [1841]).

Macquarrie, John *20th Century Religious Thought* (Philadelphia: Trinity Press International, 1989).

Marcel, Gabriel *Man Against Mass Society* trans. by G. S Fraser (Chicago: Gateway, 1962).

Marcuse, Herbert *One-Dimensional Man: Studies in the Ideology of Advanced Industrial Society* (Boston: Beacon Press, 1964).

__________., *Eros and Civilization: A Philosophical Inquiry to Freud* (New York: Vintage, 1955).

Matson, Floyd W. *The Broken Image: Man, Science and Society* (Garden City, NY: Anchor Books, 1964).

McClellan, James E. III and Harold Dorn, *Science and Technology in World History*, 2nd Edition (Baltimore, MD: The Johns Hopkins University Press, 2006).

McCurdy, Howard E. *Space and the American Imagination* (Washington DC: Smithsonian Institution, Press, 1997).

McDougall, Walter A. *The Heavens and the Earth: A Political History of the Space Age* (New York: Basic Books, 1985).

Menville, Douglas and R. Reginald, *Things to Come: An Illustrated History of the Science Fiction Film* (New York: New York Times Books, 1977).

Midgley, Mary *Science as Salvation: A Modern Myth and Its Meaning* (New York: Routledge, 1992).

Mumford, Lewis *The Myth of the Machine: Technics and Human Development* (New York: HBJ, 1966).

__________., *The Pentagon of Power: The Myth of the Machine*, Volume Two (New York: Harvest, 1970).

__________., *The Transformations of Man* (New York: Collier Books, 1966).

Nef, John U. *War and Human Progress: An Essay on the Rise of Industrial Civilization* (New York: Norton, 1968).

Nietzsche, Fredrich *Thus Spoke Zarathustra: A Book for None and All* trans., by Walter Kaufmann (New York: Penguin, 1966 [1891]).

__________., *The Gay Science* (New York: Vintage, 1974 [1882]).

__________., *On the Advantage and Disadvantage of History for Life* trans., by Peter Preuss (Indianapolis: Hackett, 1980 [1874]).

Nelson, Thomas A. *Kubrick: Inside a Film Artist's Maze* (Bloomington, IN: Indiana University Press, 2000).

Niebuhr, Reinhold *Moral Man in Immoral Society: A Study in Ethics and Politics* (New York: Charles Scribner's Sons 1932).

Noble, David F. *The Religion of Technology: The Divinity of Man and the Spirit of Invention* (New York: Knopf, 1997).

O'Connell, Mark *To Be A Machine: Adventures Among Cyborgs, Utopians, Hackers and the Futurists Solving the Modest Problem of Death* (New York: Doubleday, 2017).

Orwell, George *1984* (New York: HBC, 1949).

Otto, Rudolf *The Idea of the Holy* (New York: Oxford University Press, 1958).

Pacey, Arnold *The Maze of Ingenuity: Ideas and Idealism in the Development of Technology* 2nd Ed., (Cambridge, MA: MIT Press, 1992).

Parish, James R. and Michael R. Pitts, *The Great Science Fiction Pictures* (Metuchen, NJ: The Scarecrow Press, 1977).

Polak, Fred *The Image of the Future*, trans., by Elise Boulding (San Francisco: Jossey-Bass Inc., 1973).

Rahner, Karl, *Dictionary of Theology* New Rev. Ed. (New York: Crossroads, 1981).

Rainbow, Paul Ed., *The Foucault Reader* (New York: Pantheon, 1984).

Randell, John Herman Jr., *The Making of the Modern Mind: A Survey of the Intellectual Background of the Present Age* (New York: Columbia University Press, 1976).

Rosen, Jeffry *The Naked Crowd: Reclaiming Security and Freedom in an Anxious Age* (New York: Random House, 2004).

Ross, Richard *Waiting for the End* (Princeton, NJ: Architectural Press, 2004).

Rousas, John Rushdoony *The One and the Many: Studies in the Philosophy of Order and Ultimacy* (Fairfax: VA: Thoburn Press, 1978).

Bertrand Russell, *The Future of Science* (New York: Wisdom Library, 1959).

Sagan, Carl *Cosmos* (New York: Random House, 1980).

Sartre, Jean Paul *The Reprieve* trans., Eric Sutton (New York: Knopf, 1947).

__________., *Being and Nothingness* (New York: Washington Square Press, 1956).

Schell, Jonathan *The Fate of the Earth* (New York: Avon, 1982).

Seidenberg, Roderick *Posthistoric Man: An Inquiry* (Chapel Hill, NC: The University of North Carolina Press, 1950).

Shedd, William G. T. *Dogmatic Theology* 3rd Edition (Phillipsburg, NJ: Presbyterian and Reformed Pub., 2003 [1888]).

Sherry, Michael S. *The Rise of American Air Power: The Creation of Armageddon* (New Haven, CT: Yale University Press, 1987).

Smith, Merritt Roe and Leo Marx, Eds., *Does Technology Drive History? The Dilemma of Technological Determinism* (Cambridge, MA: The MIT Press, 1996).

Sobchack, Vivian C. *The Limits of Infinity: The American Science Fiction Film*: 1950-1975 (New York: A. S. Barnes and Company, 1980).

Sontag, Susan *On Photography* (New York: Farrar, Straus & Giroux, 1973).

Spengler, Oswald *The Decline of the West*, Two Vols, trans., by Charles Francis Atkins (New York: Knopf, 1938).

Stock, Gregory *Metaman: The Merging of Humans and Machines into a Global Superorganism* (New York: Simon and Schuster, 1993).

Taylor, Charles *A Secular Age* (Cambridge: MA: Harvard University Press, 2007).

Teich, Albert H. Ed., *Technology and the Future*, 6th Edition (New York: St. Martin's Press, 1992).

Terlizzese, Lawrence J. *Into the Void: the Coming Transhuman Transformation* (Cambridge OH: Christian Publishing House, 2016).

__________., *Hope in the Thought of Jacques Ellul* (Eugene, OR: Cascade, 2005).

__________., *Trajectory of the 21ST Century: Essays on Theology and Technology* (Eugene, OR: Cascade, 2009).

__________., *Killer Computers: Science Fiction Anticipates Our Future* (Cambridge, OH: Christian Publishing House, 2019).

Thielicke, Helmut *Modern Faith and Thought* trans., by Geoffrey W. Bromiley (Grand Rapids, MI: Eerdmans, 1990).

Tillich, Paul *The Future of Religions* (New York: Harper and Row, 1966).

__________., *The Irrelevance and Relevance of the Christian Message* (Cleveland, OH: The Pilgrim Press, 1996).

__________., *Systematic Theology*, Volume One (Chicago: The University of Chicago Press, 1951).

Tocqueville, Alexis de *Democracy in America* trans., by Henry Revee (New York: Bantam, 2000 [1835]).

Toffler, Alvin *Future Shock* (New York: Bantam, 1970).

Vahanian, Gabriel *The Death of God: The Culture of Our Post Christian Era* (New York: George Braziller, 1961).

Volti, Rudi *Society and Technological Change* 4th Edition (New York: Worth Publishers, 2001).

Wager, W. Warren *The City of Man: Prophecies of a World Civilization in Twentieth-Century Thought* (Cambridge, MA: The Riverside Press, 1963).

Watts, Alan W. Myth and Ritual in Christianity (Boston: Beacon Press, 1968).

Warren, Bill *Keep Watching the Skies!* (Jefferson: NC: McFarland, 1982).

Wiener, Philip P. Ed., *The Dictionary of the History of Ideas*, vol., III (New York: Charles Scribner's Sons, 1973).

Whitaker, Reg *The End of Privacy: How Total Surveillance is Becoming a Reality* (New York: New Press, 1999).

Woodward, Kathleen Ed., *The Myths of Information: Technology and Postindustrial* Culture (London, UK: Routledge & Kegan Paul, 1980).

Ziolkowski, Theodore *The Sin of Knowledge: Ancient Themes and Modern Variations* (Princeton: Princeton University Press, 2000).

Made in the USA
Monee, IL
04 November 2026

67379453R00105